Microgrids and Active Distribution Networks

Microgrids and Active Distribution Networks

S. Chowdhury, S.P. Chowdhury and P. Crossley

The Institution of Engineering and Technology

Contents

Foreword

Another book on distributed generation? Too much seems to have been published recently on renewable energy sources and active distribution networks. Whole conferences are devoted to the topics. Well-funded research programmes and government-regulated subsidies stimulate and encourage investment in the new systems. It is precisely this volume of material that makes this book welcome at this time.

Brought together by their common interest and complementary experience, Sunetra Chowdhury, SP Chowdhury and Peter Crossley have collated and integrated the many parallel developments in distributed generation during the past 10 years. Their book is a worthy successor to *Embedded Generation* by Nick Jenkins and his colleagues published in 2000, also by the IET. Without repeating the material in that work, the new book brings the technology up to date.

The authors have not neglected the aspects of new energy technologies that make distributed generation so attractive from an environmental perspective. However, the emphasis in this book is on the supporting technologies that can turn conventional passive electricity delivery networks into the active networks of the future. The focus is clearly on integrating the new, dispersed sources with the systems of dominant central generation, and on allowing the new technologies to operate effectively in isolated systems.

Power electronics, communications and protection introduce the 'smart' into the grids of the future. The point is often lost that smart grids can be no better than the smart engineers who conceive and implement them. The clear presentation of the details of these enabling technologies makes this book a valuable reference for research students and those engineers involved in the planning, design and installation of new systems or the upgrading of existing ones.

But this book is not only for the technical specialists. Technology in isolation is seldom successful. Good designs succeed in the appropriate context. So this book also presents the economic and market aspects of active systems and dispersed generation, which are needed to complement the technical expertise of the engineers.

Finally, this book comes at a critical time in the environmental and economical evolution of the electricity industry. For many already in the industry and the students who will be entering it, the future will be different, confusing, challenging and even exciting. A sound understanding of the material in this book will help individuals and utilities adapt to the changes. Yes, it is another book on distributed generation – a timely, relevant, balanced and readable one.

Prof. CT Gaunt
University of Cape Town

Preface

Power and energy engineers, academics, researchers and stakeholders everywhere are pondering the problems of depletion of fossil fuel resources, poor energy efficiency and environmental pollution. Hence there is a new trend of generating energy locally at distribution voltage level by using small-scale, low-carbon, non-conventional and/or renewable energy sources, like natural gas, biogas, wind power, solar photovoltaic, fuel cells, microturbines, Stirling engines, etc., and their integration into the utility distribution network. This is termed as dispersed or distributed generation (DG) and the generators are termed as distributed energy resources (DERs) or microsources. In the late 1990s, the major issues related to DG were extensively investigated by the working groups of CIGRE and CIRED in their review reports. As part of the Kyoto Protocol, many countries are planning to cut down greenhouse gas emissions (carbon and nitrogen by-products) to counter climate change and global warming. Hence many governments are coming up with new energy generation and utilisation policies to support proper utilisation of these low-carbon generation technologies.

Conventional electricity networks are in the era of major transition from passive distribution networks with unidirectional electricity transportation to active distribution networks with DERs and hence bidirectional electricity transportation. Active distribution networks need to incorporate flexible and intelligent control systems in order to harness clean energy from renewable DERs. They should also employ future network technologies for integration of DERs as smartgrid or Microgrid networks. The present 'fit-and-forget' strategy of DER deployment must be changed in active network management for accommodating a high degree of DG penetration. For actually implementing Microgrids and active distribution networks on a commercial basis, extensive research is needed, but not restricted to the following areas: (i) wide area active control, (ii) adaptive protection and control, (iii) network management devices, (iv) real-time network simulation, (v) advanced sensors and measurements, (vi) distributed pervasive communication, (vii) knowledge extraction by intelligent methods and (viii) novel design of transmission and distribution systems.

To the best of our knowledge, this book is the first of its kind to deal with various technical and economical aspects and issues of Microgrids and active distribution networks. Microgrids, as active low- and medium-voltage networks, can potentially provide a huge benefit to the main power utility by improving its energy efficiency, power quality and reliability to customers'

Abbreviations

AC	alternating current
ACC	annual capacity charge
ADS	automated dispatch system
AGC	automatic generation control
AI	artificial intelligence
ATC	available transfer capability
ATS	automatic transfer switch
BBS	Bricks-Buses-Software
BCCHP	building combined cooling, heating and power
BCU	bay control unit
BMU	bay monitoring unit
BPU	bay protection unit
CAD	computer aided design
CAN	control area network
CB	circuit breaker
CBM	capacity benefit margin
CC	central controller
CCGT	combined cycle gas turbine
CCS	carbon capture and storage
CdTe	cadmium telluride
CERTS	Consortium for Reliability Technology Solutions
CHP	combined heat and power
CIGRE	International Council on Large Electric Systems
CIRED	International Conference and Exhibition on Electricity Distribution
CIS	copper indium diselenide
CMD	converter manufacturing data
CO	carbon monoxide
COP	coefficient of performance
CT	current transformer
DA	distribution automation
DBMS	database management system
DC	direct current
DCS	distributed control system
DER	distributed energy resource

DES	district energy systems
DFIG	doubly fed induction generator
DG	distributed generation
DGCG	distributed generation coordination group
DISCOs	distribution companies
DMS	distribution management systems
DNO	distribution network operator
DO	digital output
DOE	department of energy
DPG	Data Processing Gateway
DRAM	dynamic read write memory
DSM	demand side management
DTI	Department of Trade and Industry
DUoS	distribution use of system
DVR	dynamic voltage restorer
EPROM	erasable programmable read-only memory
EMI	electromagnetic interference
EMM	Energy Management Module
EMS	energy management systems
ESQCR	Electricity Safety, Quality and Continuity Regulations
FERC	Federal Energy Regulatory Commission
FLC	fuzzy logic controller
FPGA	field programmable gate array
GENCOs	generation companies
GHG	greenhouse gas
GOOSE	Generic Object-Oriented Substation Event
GSE	Generic Substation Event
GSSE	Generic Substation Status Event
GT	gas turbine
HHI	Herfindahl–Hirschman Index
HIT	heterojunction with intrinsic thin layer
HJ	heterojunction
HMI	human–machine interface
HVAC	heat ventilation air conditioning
IC	internal combustion
IEC	International Electrotechnical Committee
IED	intelligent electronic device
IGBT	insulated gate bipolar transistors
IMDS	Information Monitoring and Diagnostic System
I/O	input/output
IPM	integrated power modules
IPP	Independent Power Producer
ISO	independent system operator
IT	information technology
LAN	local area network

LMP	locational marginal price
LN	logical node
LOG	loss of grid
LP	linear programming
LV	low voltage
MC	microsource controller
MCFC	molten carbonate fuel cell
MCP	market clearing price
MMI	man–machine interface
MOV	metal oxide varistor
MPPT	maximum power point tracking
MT	microturbine
MTG	microturbine-generator
MV	medium voltage
NEC	National Electric Code
NEREC	National Education Research and Evaluation Center
NFFO	Non-Fossil Fuel Obligation
NLP	non-linear programming
OASIS	Open Access Same-Time Information System
Ofgem	Office of Gas and Electricity Markets
OSI	Open System International
PAFC	phosphoric acid fuel cell
PC	personal computer
PCB	printed circuit board
PCC	point of common coupling
PCM	Protection Coordination Module
PCU	power-conditioning unit
PEBB	power electronic building blocks
PEI	power electronic interface
PEMFC	proton exchange membrane fuel cell
P-f	power-frequency
PI	proportional-integral
PID	proportional-integral-derivative
PLC	programmable logic controller
PMSG	permanent magnet synchronous generator
PSERC	Power System Engineering Research Center-Wisconsin
PQ	power quality
PT	potential transformer
PV	photovoltaic
PWM	pulse-width-modulated
PX	power exchange
RCC	remote control centre
RES	Renewable energy source
RIG	Remote Intelligent Gateway
RMS	root mean square

ROC	Renewable Obligation Certificate
RTP	real-time price
RTU	remote terminal unit
SC	scheduling coordinator
SCADA	supervisory control and data acquisition
SCR	silicon-controlled rectifier
SCU	station control unit
SFC	Sequential Function Charts
SG	synchronous generator
SLA	Service Level Agreement
SOFC	solid oxide fuel cell
SONET	synchronous optical networking
SSTS	static source transfer switch
STATCOM	static synchronous compensator
STC	Standard Test Conditions
SVC	static VAR Compensator
T&D	transmission and distribution
TC	Technical Committee
TCP/IP	Transmission Control Protocol/Internet Protocol
TCR	Thyristor controlled reactor
TES	thermal energy storage
THC	total hydrocarbons
TOU	time of use
TRANSCOs	transmission companies
TRM	transmission reliability margin
TSC	Thyristor switched capacitor
TSR	tip speed ratio
TTC	total transfer capability
TVSS	transient voltage surge suppressors
UCA	Utility Communications Architecture
UPS	uninterrupted power supply
VLSI	very large scale integration
VPN	virtual private network
VTB	virtual test bed
WDI	wet diluent injection
WECC	Western Electricity Coordinating Council
WECS	wind energy conversion systems
WisPERC	Wisconsin Power Electronics Research Center
WSCC	Western System Coordinating Council

1.2   Why integration of distributed generation?

In spite of several advantages provided by conventional power systems, the following technical, economic and environmental benefits have led to gradual development and integration of DG systems:

(1) Due to rapid load growth, the need for augmentation of conventional generation brings about a continuous depletion of fossil fuel reserve. Therefore, most of the countries are looking for non-conventional/renewable energy resources as an alternative.
(2) Reduction of environmental pollution and global warming acts as a key factor in preferring renewable resources over fossil fuels. As part of the Kyoto Protocol, the EU, the UK and many other countries are planning to cut down greenhouse gas (carbon and nitrogenous by-products) emissions in order to counter climate change and global warming. Therefore, they are working on new energy generation and utilisation policies to support proper utilisation of these energy sources. It is expected that exploitation of DERs would help to generate eco-friendly clean power with much lesser environmental impact.
(3) DG provides better scope for setting up co-generation, trigeneration or CHP plants for utilising the waste heat for industrial/domestic/commercial applications. This increases the overall energy efficiency of the plant and also reduces thermal pollution of the environment.
(4) Due to lower energy density and dependence on geographical conditions of a region, DERs are generally modular units of small capacity. These are geographically widespread and usually located close to loads. This is required for technical and economic viability of the plants. For example, CHP plants must be placed very close to their heat loads, as transporting waste heat over long distances is not economical. This makes it easier to find sites for them and helps to lower construction time and capital investment. Physical proximity of load and source also reduces the transmission and distribution (T&D) losses. Since power is generated at low voltage (LV), it is possible to connect a DER separately to the utility distribution network or they may be interconnected in the form of Microgrids. The Microgrid can again be connected to the utility as a separate semi-autonomous entity.
(5) Stand-alone and grid-connected operations of DERs help in generation augmentation, thereby improving overall power quality and reliability. Moreover, a deregulated environment and open access to the distribution network also provide greater opportunities for DG integration. In some countries, the fuel diversity offered by DG is considered valuable, while in some developing countries, the shortage of power is so acute that any form of generation is encouraged to meet the load demand.

1.3   Active distribution network

Electricity networks are in the era of major transition from stable passive distribution networks with unidirectional electricity transportation to active distribution networks with bidirectional electricity transportation. Distribution networks without any DG units are passive since the electrical power is supplied by the national grid system to the customers embedded in the distribution networks. It becomes active when DG units are added to the distribution system leading to bidirectional power flows in the networks. To effect this transition, developing countries should emphasise the development of sustainable electricity infrastructure while the developed countries should take up the technical and economic challenges for the transformation of distribution networks. The UK industry regulator, the Office of Gas and Electricity Markets (Ofgem), has named this challenge as 'Rewiring Britain'. Active distribution networks need to incorporate flexible and intelligent control with distributed intelligent systems. In order to harness clean energy from renewable DERs, active distribution networks should also employ future network technologies leading to smartgrid or Microgrid networks.

Present 'fit-and-forget' strategy of DG employment needs to be changed in active network management. It should incorporate integration of DGs in distribution networks and demand side management. It has been demonstrated by the UK-based Centre for Distributed Generation and Sustainable Electrical Energy (www.sedg.ac.uk) that the application of active network management methods can greatly support more DG connections as compared to networks without active management.

Several Department of Trade and Industry (DTI) and Ofgem reports clearly indicate that intelligent active distribution networks have gained momentum. Several factors are in favour of the evolution of active distribution networks, e.g. (i) pressing customer expectations of high-quality reliable power distribution, (ii) increasing desire of policy makers for accommodation of renewable DERs with energy storage devices, (iii) carbon commitment in reducing emissions by 50% by 2050, (iv) motivating the distribution network operators (DNOs) towards better asset utilisation and management by deferral of replacement of age-old assets, etc.

In order to implement evolutionary active distribution networks for flexible and intelligent operation and control, extensive research is necessary. The focus of the research should be mainly in the following areas: (i) wide area active control, (ii) adaptive protection and control, (iii) network management devices, (iv) real-time network simulation, (v) advanced sensors and measurements, (vi) distributed pervasive communication, (vii) knowledge extraction by intelligent methods and (viii) novel design of transmission and distribution systems.

1.4   Concept of Microgrid

Microgrids are small-scale, LV CHP supply networks designed to supply electrical and heat loads for a small community, such as a housing estate or a suburban locality, or an academic or public community such as a university or school, a commercial area, an industrial site, a trading estate or a municipal region. Microgrid is essentially an active distribution network because it is the conglomerate of DG systems and different loads at distribution voltage level. The generators or microsources employed in a Microgrid are usually renewable/non-conventional

DERs integrated together to generate power at distribution voltage. From operational point of view, the microsources must be equipped with power electronic interfaces (PEIs) and controls to provide the required flexibility to ensure operation as a single aggregated system and to maintain the specified power quality and energy output. This control flexibility would allow the Microgrid to present itself to the main utility power system as a single controlled unit that meets local energy needs for reliability and security.

The key differences between a Microgrid and a conventional power plant are as follows:

(1) Microsources are of much smaller capacity with respect to the large generators in conventional power plants.
(2) Power generated at distribution voltage can be directly fed to the utility distribution network.
(3) Microsources are normally installed close to the customers' premises so that the electrical/heat loads can be efficiently supplied with satisfactory voltage and frequency profile and negligible line losses.

The technical features of a Microgrid make it suitable for supplying power to remote areas of a country where supply from the national grid system is either difficult to avail due to the topology or frequently disrupted due to severe climatic conditions or man-made disturbances.

From grid point of view, the main advantage of a Microgrid is that it is treated as a controlled entity within the power system. It can be operated as a single aggregated load. This ascertains its easy controllability and compliance with grid rules and regulations without hampering the reliability and security of the power utility. From customers' point of view, Microgrids are beneficial for locally meeting their electrical/heat requirements. They can supply uninterruptible power, improve local reliability, reduce feeder losses and provide local voltage support. From environmental point of view, Microgrids reduce environmental pollution and global warming through utilisation of low-carbon technology.

However, to achieve a stable and secure operation, a number of technical, regulatory and economic issues have to be resolved before Microgrids can become commonplace. Some problem areas that would require due attention are the intermittent and climate-dependent nature of generation of the DERs, low energy content of the fuels and lack of standards and regulations for operating the Microgrids in synchronism with the power utility. The study of such issues would require extensive real-time and off line research, which can be taken up by the leading engineering and research institutes across the globe.

1.5   A typical Microgrid configuration

A typical Microgrid configuration is shown in Figure 1.1. It consists of electrical/heat loads and microsources connected through an LV distribution network. The loads (especially the heat loads) and the sources are placed close together to

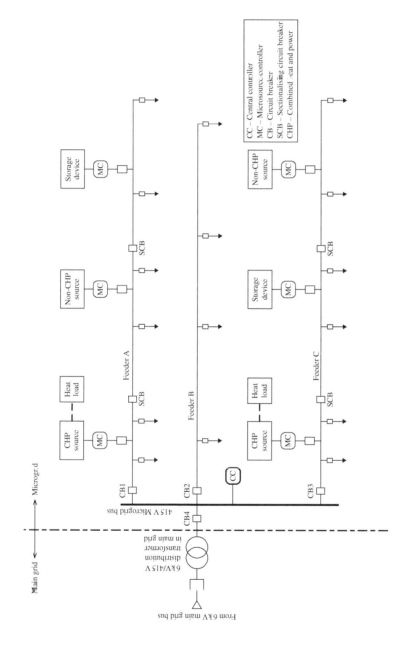

CC – Central controller
MC – Microsource controller
CB – Circuit breaker
SCB – Sectionalising circuit breaker
CHP – Combined heat and power

Figure 1.1   A typical Microgrid configuration

protecting the sensitive loads. PCM also helps to re-synchronise the Microgrid to the main grid after the initiation of switchover to the grid-connected mode of operation through suitable reclosing schemes.

The functions of the CC in the grid-connected mode are as follows:

(1) Monitoring system diagnostics by collecting information from the micro-sources and loads.
(2) Performing state estimation and security assessment evaluation, economic generation scheduling and active and reactive power control of the microsources and demand side management functions by using collected information.
(3) Ensuring synchronised operation with the main grid maintaining the power exchange at priori contract points.

The functions of the CC in the stand-alone mode are as follows:

(1) Performing active and reactive power control of the microsources in order to maintain stable voltage and frequency at load ends.
(2) Adopting load interruption/load shedding strategies using demand side management with storage device support for maintaining power balance and bus voltage.
(3) Initiating a local black start to ensure improved reliability and continuity of service.
(4) Switching over the Microgrid to grid-connected mode after main grid supply is restored without hampering the stability of either grid.

1.6   Interconnection of Microgrids

Since Microgrids are designed to generate power at distribution voltage level along with utilisation of waste heat, they have restricted energy handling capability. Therefore, their maximum capacity is normally restricted to approximately 10 MVA as per IEEE recommendations. Hence, it is possible to supply a large load pocket from several Microgrids through a common distribution network, by splitting the load pocket into several controllable load units, with each unit being supplied by one Microgrid. In this way, Microgrids can be interconnected to form much larger power pools for meeting bulk power demands. For interconnected Microgrids, each CC must execute its control in close co-ordination with the neighbouring CCs. Thus, an interconnected Microgrid would achieve greater stability and controllability with a distributed control structure. It would also have more redundancy to ensure better supply reliability.

1.7   Technical and economical advantages of Microgrid

The development of Microgrid is very promising for the electric energy industry because of the following advantages:

(1) *Environmental issues* – It is needless to say that Microgrids would have much lesser environmental impact than the large conventional thermal power stations. However, it must be mentioned that the successful implementation of carbon capture and storage (CCS) schemes for thermal power plants will drastically reduce the environmental impacts. Nevertheless, some of the benefits of Microgrid in this regard are as follows:
  (i) Reduction in gaseous and particulate emissions due to close control of the combustion process may ultimately help combat global warming.
  (ii) Physical proximity of customers with microsources may help to increase the awareness of customers towards judicious energy usage.

(2) *Operation and investment issues* – Reduction of physical and electrical distance between microsource and loads can contribute to:
  (i) Improvement of reactive support of the whole system, thus enhancing the voltage profile.
  (ii) Reduction of T&D feeder congestion.
  (iii) Reduction of T&D losses to about 3%.
  (iv) Reduction/postponement of investments in the expansion of transmission and generation systems by proper asset management.

(3) *Power quality* – Improvement in power quality and reliability is achieved due to:
  (i) Decentralisation of supply.
  (ii) Better match of supply and demand.
  (iii) Reduction of the impact of large-scale transmission and generation outages.
  (iv) Minimisation of downtimes and enhancement of the restoration process through black start operations of microsources.

(4) *Cost saving* – The following cost savings are achieved in Microgrid:
  (i) A significant saving comes from utilisation of waste heat in CHP mode of operation. Moreover, as the CHP sources are located close to the customer loads, no substantial infrastructure is required for heat transmission. This gives a total energy efficiency of more than 80% as compared to a maximum of 40% for a conventional power system.
  (ii) Cost saving is also effected through integration of several microsources. As they are locally placed in plug-and-play mode, the T&D costs are drastically reduced or eliminated. When combined into a Microgrid, the generated electricity can be shared locally among the customers, which again reduces the need to import/export power to/from the main grid over longer feeders.

(5) *Market issues* – The following advantages are attained in case of market participation:
  (i) The development of market-driven operation procedures of the Microgrids will lead to a significant reduction of market power exerted by the established generation companies.
  (ii) The Microgrids may be used to provide ancillary services.

(iii) Widespread application of modular plug-and-play microsources may contribute to a reduction in energy price in the power market.

(iv) The appropriate economic balance between network investment and DG utilisation is likely to reduce the long-term electricity customer prices by about 10%.

1.8   Challenges and disadvantages of Microgrid development

In spite of potential benefits, development of Microgrids suffers from several challenges and potential drawbacks as explained.

(1) *High costs of distributed energy resources* – The high installation cost for Microgrids is a great disadvantage. This can be reduced by arranging some form of subsidies from government bodies to encourage investments. This should be done at least for a transitory period for meeting up environmental and carbon capture goals. There is a global target set to enhance renewable green power generation to 20% by 2020 and to reduce carbon emission by 50% by 2050.

(2) *Technical difficulties* – These are related to the lack of technical experience in controlling a large number of plug-and-play microsources. This aspect requires extensive real-time and off line research on management, protection and control aspects of Microgrids and also on the choice, sizing and placement of microsources. Specific telecommunication infrastructures and communication protocols must be developed in this area. Research is going on for the implementation and roll-out of IEC 61850 in communication for Microgrid and active distribution networks. However, lack of proper communication infrastructure in rural areas is a potential drawback in the implementation of rural Microgrids. Besides, economic implementation of seamless switching between operating modes is still a major challenge since the available solutions for reclosing adaptive protection with synchronism check are quite expensive.

(3) *Absence of standards* – Since Microgrid is a comparatively new area, standards are not yet available for addressing operation and protection issues. Power quality data for different types of sources, standards and protocols for integration of microsources and their participation in conventional and deregulated power markets, safety and protection guidelines, etc., should be laid down. Standards like G59/1 and IEEE 1547 should be reassessed and restructured for the successful implementation of Microgrid and active distribution networks.

(4) *Administrative and legal barriers* – In most countries, no standard legislation and regulations are available to regulate the operation of Microgrids. Governments of some countries are encouraging the establishment of green power Microgrids, but standard regulations are yet to be framed for implementation in future.

(5) *Market monopoly* – If the Microgrids are allowed to supply energy autonomously to priority loads during any main grid contingency, the main question that arises is who will then control energy supply prices during the period over

which main grid is not available. Since the main grid will be disconnected and the current electricity market will lose its control on the energy price, Microgrids might retail energy at a very high price exploiting market monopoly. Thus, suitable market infrastructure needs to be designed and implemented for sustaining development of Microgrids.

1.9 Management and operational issues of a Microgrid

Major management and operational issues related to a Microgrid are as follows:

(1) For maintaining power quality, active and reactive power balance must be maintained within the Microgrid on a short-term basis.
(2) A Microgrid should operate stand-alone in regions where utility supply is not available or in grid-connected mode within a larger utility distribution network. Microgrid operator should be able to choose the mode of operation within proper regulatory framework.
(3) Generation, supply and storage of energy must be suitably planned with respect to load demand on the Microgrid and long-term energy balance.
(4) Supervisory control and data acquisition (SCADA) based metering, control and protection functions should be incorporated in the Microgrid CCs and MCs. Provisions must be made for system diagnostics through state estimation functions.
(5) Economic operation should be ensured through generation scheduling, economic load dispatch and optimal power flow operations.
(6) System security must be maintained through contingency analysis and emergency operations (like demand side management, load shedding, islanding or shutdown of any unit). Under contingency conditions, economic rescheduling of generation should be done to take care of system loading and load-end voltage/frequency.
(7) Temporary mismatch between generation and load should be alleviated through proper load forecasting and demand side management. The shifting of loads might help to flatten the demand curve and hence to reduce storage capacity.
(8) Suitable telecommunication infrastructures and communication protocols must be employed for overall energy management, protection and control. Carrier communication and IEC 61850 communication infrastructures are most likely to be employed.

1.10 Dynamic interactions of Microgrid with main grid

The capacity of Microgrid being sufficiently small, the stability of main grid is not affected when it is connected to the main grid. However, in future, when Microgrids will become more commonplace with higher penetration of DERs, the stability and security of the main grid will be influenced significantly. In such case,

Chapter 2
Distributed energy resources

2.1 Introduction

Renewable or non-conventional electricity generators employed in DG systems or Microgrids are known as distributed energy resources (DERs) or microsources. One major aim of Microgrids is to combine all benefits of non-conventional/renewable low-carbon generation technologies and high-efficiency combined heat and power (CHP) systems. In this regard, the CHP-based DERs facilitate energy-efficient power generation by capturing waste heat while low-carbon DERs help to reduce environmental pollution by generating clean power. Prospective DERs range from micro-CHP systems based on Stirling engines, fuel cells and micro-turbines to renewables like solar photovoltaic (PV) systems, wind energy conversion systems (WECS) and small-scale hydroelectric generation. Choice of a DER very much depends on the climate and topology of the region and fuel availability. Possibilities of using biofuels and application of various storage technologies like flywheel batteries and ultracapacitors are also being investigated across the globe in the field of Microgrid research. Most of the countries are coming up with schemes to support the exploitation of the renewable/non-conventional energy resources for meeting up global carbon commitment.

This chapter briefly describes the following DER technologies:

- Combined heat and power (CHP) systems
- Wind energy conversion systems (WECS)
- Solar photovoltaic (PV) systems
- Small-scale hydroelectric generation
- Other renewable energy sources
- Storage devices.

2.2 Combined heat and power (CHP) systems

CHP or cogeneration systems are most promising as DERs for Microgrid applications. Their main advantage is energy-efficient power generation by judicious utilisation of waste heat. Unlike fossil-fuelled power plants, CHP systems capture and use the by-product heat locally for domestic and industrial/process heating purposes. Heat produced at moderate temperatures (100–180 °C) can also be used in absorption chillers for cooling. Simultaneous production of electricity, heat and cooling is known as trigeneration or polygeneration.

By capturing the excess heat, CHP system allows better usage of energy than conventional generation, potentially reaching an efficiency of more than 80%, compared with that of about 35% for conventional power plants. It is most efficient when the heat is utilised locally. Overall efficiency is reduced if heat is to be transported over long distances using heavily insulated pipes, which are both expensive and inefficient. On the other hand, electricity can be transmitted over much longer distances for lesser energy loss. Thus, CHP plants can be located somewhat remotely from their electrical loads, but they must always be located close to the heat loads for better performance. CHP plants are commonly employed in district heating systems of big towns, hospitals, prisons, oil refineries, paper mills and industrial plants with large heat loads.

Use of CHP plants has been found to lead to 35% reduction in primary energy use as compared to conventional power generation and heat-only boilers, 30% reduction in emission with respect to coal-fired power plants and 10% reduction in emission with respect to combined cycle gas-turbine plants.

2.2.1   Micro-CHP systems

Micro-CHP systems are usually installed in smaller premises like homes or small commercial buildings. They differ from larger CHP units not only in terms of their energy-producing capacities but also in matters of parameter-driven operation. Most large industrial CHP units generate electricity as the primary product with heat as secondary while micro-CHP systems generate heat as the primary commodity with electricity as a by-product. Thus, energy generation of micro-CHP systems is mainly dictated by the heat demand of the end-users. Because of this operating model and the fluctuating electrical demand of the structures they operate in (like homes and small commercial buildings), micro-CHP systems often generate more electricity than is demanded.

Micro-CHP sets are basically microturbines coupled to single-shaft, high-speed (50,000–100,000 rpm) permanent magnet synchronous machines with airfoil or magnetic bearings. They are provided with power electronic interfaces for connection to the electrical loads. They also have their own heat recovery systems for low and medium temperature heat extraction. Micro-CHP sets are reliable, robust and cheap. They are available in the range of 10–100 kW capacity. The primary fuel is natural gas, propane or liquid fuel, which permits clean combustion with low particulates. Biofuelled microturbines are also being considered as a possibility.

During operation of a CHP set, the pressure of incoming air is raised after passing through the centrifugal compressor. Temperature of the compressed air is increased on passing through the heat exchanger. When the hot compressed air enters the combustion chamber, it is mixed with fuel and burnt. The high-temperature combustion gases are expanded in the turbine to produce mechanical power, which in turn drives the permanent magnet synchronous machine to produce electrical power at high frequency. High-frequency output voltage is converted into

DC using a rectifier and the DC voltage is re-converted into AC of 50/60 Hz of frequency as per necessity using an inverter interface.

Microgrids can secure the following major advantages by using micro-CHP plants:

(1) Since transportation of electricity is far easier and more cost-effective than that of heat, it is much more suitable to place micro-CHP plants near heat loads than electrical loads. Microgrid permits this energy optimal placement of CHP plants to achieve full utilisation of heat. In case of necessity, fuel cells can also be used in the CHP plants for better utilisation of the generated heat.
(2) The scale of heat generation for individual units is small. Therefore, micro-CHP plants have greater flexibility in matching several small heat loads. Technically, a Microgrid can be designed with a judicious mix of waste and non-waste heat-producing generators so as to optimise the combined generation of heat and electricity. In spite of the aforesaid flexibility, chances are still there of having mismatch in generating a proper mix of heat and electricity. Hence, attention must be paid to enhance this flexibility.

Micro-CHP systems are primarily based on the following technologies:

(1) Internal combustion (IC) engines
(2) Stirling engines
(3) Microturbines
(4) Fuel cells.

2.2.1.1   Internal combustion (IC) engines

In IC engines, fuel is burnt in air in a combustion chamber with or without oxidisers. Combustion creates high-temperature and high-pressure gases that are allowed to expand and act on movable bodies like pistons or rotors. IC engines are different from external combustion engines like steam engines and Stirling engines. The external combustion engines use the combustion process to heat a separate working fluid which then works by acting on the movable parts. IC engines include intermittent combustion engines (e.g. reciprocating engine, Wankel engine and Bourke's engine) and the continuous combustion engines (e.g. Jet engines, rockets and gas turbines).

The commonly used fuels are diesel, gasoline and petroleum gas. Propane gas is also sometimes used as fuel. With some modifications to the fuel delivery components, most IC engines designed for gasoline can run on natural gas or liquefied petroleum gases. Liquid and gaseous biofuels, like ethanol and biodiesel, may also be used. Depending on the type of fuel, the IC engines are provided with spark ignition or compression ignition systems in their cylinders to initiate the fuel combustion process.

2.2.1.2   Stirling engines

Stirling engine is a closed-cycle piston heat engine where the working gas is permanently contained within the cylinder. It is traditionally classified as an external combustion engine, though heat can also be supplied by non-combustible

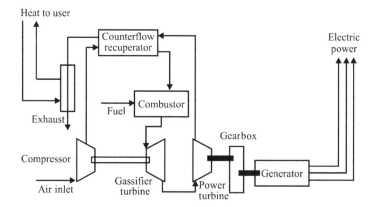

Figure 2.2   Split-shaft microturbine

(5) *Economy of operation* – System costs are lower than $500 per kW. Cost of electricity is competitive with alternatives including grid power for market applications.
(6) *Fuel flexibility* – It is capable of using alternative fuels, like natural gas, diesel, ethanol and landfill gas, and other biomass-derived liquids and gases.
(7) *Noise level* – It has reduced level of noise and vibrations.
(8) *Installation* – It has simpler installation procedure.

A survey of research literature shows that there is an extensive thrust on the application of microturbine as DG systems. Research areas include simulation, offline/real-time studies and development of inverter interfaces for microturbine applications. Several research papers are available on the development of a single-stage axial flow microturbine for power generation, on the study of the facilities of the technology through relevant test results, on the development of active filters and adaptive control mechanisms for microturbines in hybrid power systems, etc. Studies also include development of dynamic models for microturbines to analyse their performance in islanded and grid-connected modes of operation and for cogeneration applications.

Most microturbines use permanent magnet synchronous generator (PMSG) or asynchronous generator for power generation. Ample research has been conducted on PMSG-coupled microturbines. However, very little has been reported on development and load-following performance analysis of microturbine models with synchronous generator (SG) in islanded and grid-connected modes. This area needs to be extensively investigated to resolve the technical issues for integrated operation of a microturbine with the main utility grid.

The main advantage of coupling an SG with a split-shaft microturbine is that it eliminates the use of the power converter. In this case, the generator is connected to the turbine via a gearbox to generate conventional 50/60 Hz power. Thus, the need for rectifiers and power converter units is completely eliminated. Moreover, the use of high-speed PMSG has disadvantages – such as thermal stress, demagnetisation phenomena, centrifugal forces, rotor losses – because of fringing effects and high cost. The disadvantage of coupling induction (asynchronous) generators is that though they are cheaper and robust, their speed is load dependent and they cannot be connected to the grid without the use of expensive power converter systems. The use of power electronic interfaces for power conversion introduces harmonics in the system to reduce the output power quality. These harmonics are eliminated if an SG is used with a gearbox. Also, there are less chances of failure as the gearbox is a much simpler mechanical equipment as compared to complex power electronic devices. However, the main drawback of using a gearbox is that it consumes a fraction of generated power, thus reducing the efficiency of the system. Some manufacturing companies like Ingersoll-Rand Energy Systems, Ballard, Bowman and Elliott are using synchronous machines with their microturbines for both stand-alone and grid-connected operations. Major parts of a microturbine and their functions are as follows:

(1) *Turbine* – High-speed single-shaft or split-shaft gas turbines.
(2) *Alternator* – In single-shaft units, the alternator is directly coupled to the turbine. The rotor is either two-pole or four-pole permanent design and the stator is of conventional copper wound design. In split-shaft units, a conventional induction machine or synchronous machine is mounted on the turbine through the gearbox.
(3) *Power electronics* – In single-shaft machines, the high-frequency (1,500–4,000 Hz) AC voltage generated by the alternator is converted into standard power frequency voltage through the power electronic interfaces. However, in the split-shaft design, these are not required due to the presence of the gearbox.
(4) *Recuperator* – The recuperator recovers the waste heat to improve the energy efficiency of the microturbine. It transfers heat from the exhaust gas to the discharge air before the discharge air enters the combustor. This reduces the amount of fuel needed to raise the discharge air temperature to the required value. The process of designing and manufacturing recuperators is quite complicated as they operate under high pressure and temperature differentials. Exhaust heat can be used for water heating, drying processes or absorption chillers for air conditioning from heat energy instead of electric energy.
(5) *Control and communication* – Control and communication systems include the entire turbine control mechanism, inverter interface, start-up electronics, instrumentation and signal conditioning, data logging, diagnostics and user control communications.

2.2.1.4   Fuel cells

A fuel cell converts chemical energy of a fuel directly into electrical energy. It consists of two electrodes (an anode and a cathode) and an electrolyte, retained in a matrix. The operation is similar to that of a storage battery except that the reactants and products are not stored, but are continuously fed to the cell. During operation, the hydrogen-rich fuel and oxidant (usually air) are separately supplied to the

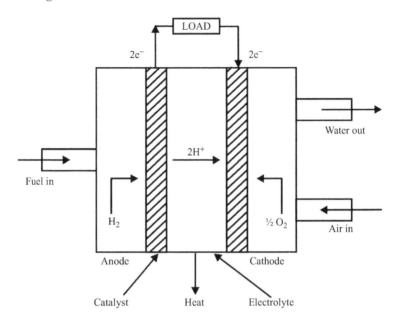

Figure 2.3   Basic construction of a fuel cell

electrodes. Fuel is fed to the anode and oxidant to the cathode, and the two streams are separated by an electrode–electrolyte system. Electrochemical oxidation and reduction take place at the electrodes to produce electricity. Heat and water are produced as by-products. Figure 2.3 shows the basic construction of a fuel cell.

Fuel cells have several advantages over conventional generators. Due to higher efficiency and lower fuel oxidation temperature, fuel cells emit less CO_2 and NO_x per kilowatt of power generated. Thus, they provide an eco-friendly energy source. As there are no moving parts, they are almost free from noise and vibration, robust and low maintenance. This makes them suitable for urban or suburban locations. Unlike gas and steam turbines, fuel cell efficiency increases at part-load conditions. Moreover, they can use a variety of fuels like natural gas, propane, landfill gas, anaerobic digester gas, diesel, naptha, methanol and hydrogen. This versatility ensures that this technology will not become obsolete due to the unavailability of fuels.

A single fuel cell produces output voltage less than 1 V. Therefore, to produce higher voltages, fuel cells are stacked on top of each other and are series connected forming a fuel cell system. Electrical efficiencies of fuel cells lie between 36% and 60%, depending on the type and system configuration. By using conventional heat recovery equipment, overall efficiency can be enhanced to about 85%.

Steam reforming of liquid hydrocarbons (C_nH_m) is a potential way of providing hydrogen-rich fuel for fuel cells. This is preferred because storage of hydrogen is quite hazardous and expensive. Reformers provide a running stream of hydrogen without having to use bulky pressurised hydrogen tanks or hydrogen vehicles for distribution. The endothermic reaction that occurs in the reforming process in the presence of a catalyst is

$$C_nH_m + nH_2O \rightarrow nCO + \left(\frac{m}{2} + n\right)H_2$$

$$CO + H_2O \rightarrow CO_2 + H_2$$

(2.1)

Carbon monoxide combines with steam to produce more hydrogen through the water gas shift reaction. Figure 2.4 shows the flows and reactions in a fuel cell.

Extensive research is going on to design reformer–fuel cell system in spite of the following challenges:

(1) Steam reforming can utilise liquid hydrocarbon fuels like ethanol and biodiesel, but these fuels may not be available in sufficiently large quantities to provide a continuous stream of hydrogen.
(2) As the reforming reaction takes place at high temperatures, fuel cells have high start-up time and require costly temperature-resistant materials.
(3) The catalyst is very expensive and the sulphur compounds present in the fuel may poison certain catalysts, making it difficult to run this type of system on ordinary gasoline.
(4) Carbon monoxide produced in the reaction may poison the fuel cell membrane and may degrade its performance. In that case, complicated CO-removal systems must be incorporated into the system.
(5) Thermodynamic efficiency of the process depends on the purity of the hydrogen product. Normally thermodynamic efficiency lies between 70% and 85%.

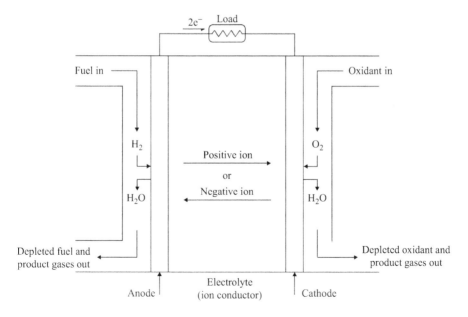

Figure 2.4   Flows and reactions in fuel cell

During operation, the oxidant enters the cathode compartment. After electrode reaction, oxygen ions migrate through the electrolyte layer to the anode where hydrogen is oxidised. The high operating temperature provides adequate heat for the endothermic reforming reaction. SOFC, thus, is more tolerant to fuel impurities and can operate using hydrogen and carbon monoxide fuels directly at the anode. It does not need external reformers or catalysts to produce hydrogen. This feature makes SOFC especially attractive for fuels like biomass or coal gasification.

When integrated with a gas turbine (SOFC-GT systems), SOFC systems achieve almost 70–75% electrical efficiencies, the highest amongst all the fuel cell technologies. They also have lifetimes of 10–20 years, which is two to four times higher than other fuel cells. SOFC is well suited for DG systems, CHP applications and applications such as small portable generators. Two different SOFC geometries are available, viz. tubular design for large capacity cogeneration and DG applications and planar design for small capacity power generation applications.

The main disadvantage of the SOFC is the stringent material requirement for the critical cell components due to high operating temperature. The use of exotic ceramics, metal–ceramic composites and high-temperature alloys and the manufacturing techniques required by these materials increase the cost of SOFCs to a great extent. Because of this, manufacturers are currently trying to reduce the operating temperature to about 700–900 °C.

2.3   Wind energy conversion systems (WECS)

WECS convert wind energy into electrical energy. The principal component of WECS is the wind turbine. This is coupled to the generator through a multiple-ratio gearbox. Usually induction generators are used in WECS. The main parts of a wind turbine are the tower, the rotor and the nacelle. The nacelle accommodates the transmission mechanisms and the generator. Rotor may have two or more blades. Wind turbine captures the kinetic energy of wind flow through rotor blades and transfers the energy to the induction generator side through the gearbox. The generator shaft is driven by the wind turbine to generate electric power. The function of the gearbox is to transform the slower rotational speeds of the wind turbine to higher rotational speeds on the induction generator side. Output voltage and frequency is maintained within specified range, by using supervisory metering, control and protection techniques. Wind turbines may have horizontal axis configuration or vertical axis configuration. The average commercial turbine size of WECS was 300 kW until the mid-1990s, but recently machines of larger capacity, up to 5 MW, have been developed and installed.

The output power of a wind turbine is determined by several factors such as wind velocity, size and shape of the turbine. The power developed is given by

$$P = \frac{1}{2} C_p \rho V^3 A \tag{2.2}$$

where P is power (W), C_p power coefficient, ρ air density (kg/m^3), V wind velocity (m/s) and A swept area of rotor blades (m^2).

Power coefficient C_p gives a measure of the amount of energy extracted by the turbine rotor. Its value varies with rotor design and the tip speed ratio (TSR). TSR is the relative speed of the rotor and the wind and has a maximum practical value of about 0.4. The torque output often suffers from dynamic variations due to fluctuations in wind speed caused by tower shadow, wind shear and turbulence. These variations lead to a dynamic perturbation in the output power and hence a flicker in the generated voltage. In a constant speed wind turbine, power variation and voltage flicker do pose a problem in the network. On the contrary, variable speed wind turbine systems provide much smoother output power and more stable bus voltage with lower losses. However, a major problem of WECS is that due to the intermittent nature of generation and energy consumption in the generating plant itself, the declared net capacity is lesser than the nameplate capacity.

2.3.1 Wind turbine operating systems

Depending on controllability, wind turbine operating systems are classified as (1) constant speed wind turbines and (2) variable speed wind turbines.

2.3.1.1 Constant speed wind turbines

These operate at almost constant speed as predetermined by the generator design and gearbox ratio. The control schemes are always aimed at maximising either energy capture by controlling the rotor torque or the power output at high winds by regulating the pitch angle. According to the control strategy, constant speed wind turbines are again subdivided into (i) stall-regulated turbines and (ii) pitch-regulated turbines.

Constant speed stall-regulated turbines have no options for any control input. Here, the turbine blades are designed with a fixed pitch to operate near the optimal TSR for a given wind speed. When wind speed increases, the angle of attack also increases. Consequently an increasingly large portion of the blade starting at the blade root enters the stall region. This results in the reduced rotor efficiency and limitation of the power output. Another variation of this concept is to operate the wind turbine at two distinct constant operating speeds by either changing the number of poles of the induction generator or changing the gear ratio. The main advantage of stall regulation is its simplicity. However, the main disadvantage is that these wind turbines are not able to capture wind energy in an efficient manner at wind speeds other than the design speed. Constant speed pitch-regulated turbines typically use pitch regulation for starting up. After start-up, power can be controlled only above the rated wind speed of the turbine.

Constant speed wind turbine operating systems have the following advantages:

(1) They have a simple, robust construction and are electrically efficient.
(2) They are highly reliable due to fewer parts.
(3) No current harmonics are generated as there is no frequency conversion.
(4) They have a lower capital cost as compared to variable speed wind turbines.

However, they have the following disadvantages as compared to variable speed turbines:

(1) They are aerodynamically less efficient.
(2) They are prone to mechanical stress and are more noisy.

2.3.1.2   Variable speed wind turbine system

A typical variable speed pitch-regulated wind turbine system is shown in Figure 2.5. It has two methods for controlling the turbine operation, viz. speed changes and blade pitch changes. The control strategies usually employed are (i) power optimisation strategy and (ii) power limitation strategy.

Power optimisation strategy is employed when the wind speed is below the rated value. This strategy optimises the energy capture by maintaining a constant speed corresponding to the optimum TSR. If, however, speed is changed due to load variation, the generator may be overloaded for wind speeds above nominal value. To avoid this, methods like generator torque control are used to control the speed. Power limitation strategy is used for wind speeds above the rated value. This strategy limits the output power to the rated value by changing the blade pitch to reduce the aerodynamic efficiency.

Variable speed wind turbine systems have the following advantages:

(1) They have high energy capture capacity and are subjected to less mechanical stress.
(2) They are aerodynamically efficient and have low transient torque.

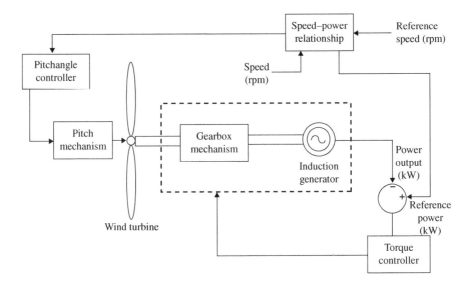

Figure 2.5   Variable speed pitch-regulated wind turbine

(3) No mechanical damping systems are required as the electrical system can effectively provide the damping.
(4) They do not suffer from synchronisation problems or voltage sags due to stiff electrical controls.

However, they have the following disadvantages as compared to constant speed turbines:

(1) They have lower electrical efficiency.
(2) They are more expensive and sometimes require complex control strategies.

2.4 Solar photovoltaic (PV) systems

Solar PV generation involves the generation of electricity from free and inexhaustible solar energy. The major advantages of a PV system are (i) sustainable nature of solar energy as fuel, (ii) minimum environmental impact, (iii) drastic reduction in customers' electricity bills due to free availability of sunlight, (iv) long functional lifetime of over 30 years with minimum maintenance and (v) silent operation. Owing to these benefits, today PV systems are recognised by governments, environmental organisations and commercial organisations as a technology with the potential to supply a significant part of the world's energy needs in a sustainable and renewable manner. Moreover, due to the extensive improvement in inverter technologies, PV generation is now being preferred and deployed worldwide as DERs for augmentation of local generation at distribution voltage level.

Though PV cells can be effectively used as a DER in a Microgrid, yet they suffer from the disadvantages of high installation cost and low energy efficiency. It has been studied that small PV installations are more cost-effective than larger ones, which indicates the effectiveness of feeding PV generation directly into customer circuits at low voltage distribution networks. However, the nature of PV generation being DC, suitable power converter circuits are to be employed for converting DC power into AC at the specified frequency level. Hence, they can be potential contributors to a Microgrid.

Solar energy reaches the PV cell in two components, direct and diffuse. The direct component is about 85% and comes through direct radiation. The diffuse component is about 15% and comes through scattered diffusion in the atmosphere. A PV cell behaves as a photodiode. Light energy incident on the cell surface in the form of photons generates electron–hole pairs as current carriers at the p-n junction. Thus, photocurrent produced by a PV cell is directly related to its surface area, incident irradiance and ambient temperature. Generated voltage is limited by the forward voltage drop across the p-n junction. As the voltage and the current output of a single cell are very small, a large number of cells are arranged in series–parallel combination to produce PV arrays or modules of higher voltage and power rating. Most PV modules are equipped with maximum power point tracking (MPPT) systems that maximise the power output from the modules by shifting the operating point depending on the solar irradiance.

Currently, water supply companies are using PV systems to trickle charge their batteries in remote monitoring equipment, while the Meteorological Office has installed PV-powered remote sensing equipment in northern UK.

2.5   Small-scale hydroelectric power generation

The small-scale hydroelectric generators are effectively used for generating power onsite in Microgrids. Extent of generation depends on the topography of an area and its annual precipitation. These generators suffer from large variations in generation due to variable water flow caused by uneven rainfall. This is particularly true for hydro power stations, which do not have their own storage reservoirs and for which the catchment area is spread over rocky soil without vegetation cover. Variable water resources also lead to varying generation with a low capacity factor. Capacity factor is defined as the ratio of available annual energy to its rated annual capacity. The power output from a hydro turbine is given by

$$P = QH\eta\rho g \tag{2.3}$$

where P is power output (W), Q water flow rate (m^3/s), H effective head (m), η overall efficiency, ρ water density (1,000 kg/m^3) and g acceleration due to gravity.

Equation (2.3) indicates that the power output can be increased by increasing both effective head and water flow rate. Penstock is the pipeline through which water is brought to the turbine. The cross section of the penstock is suitably designed for optimum water flow rate. Various types of water turbines are used depending on available water heads and flow rates. Usually, reaction turbines operate at lower heads (e.g. Francis and Kaplan turbines) and the impulse turbines (e.g. Pelton wheels and Turgo turbines) operate at higher heads. Reaction turbines extract energy from pressure drop whereas impulse turbines extract power from the kinetic energy of water jets at atmospheric pressure. Cross-flow impulse turbines are used for small hydro units where kinetic energy is extracted from water striking the turbine blades as a water sheet rather than a jet. Both synchronous and induction generators may be used for small-scale hydrogeneration with suitable multiple-ratio gearboxes. However, suitable precautions must be taken during designing a turbine so that its damage due to overspeeding can be avoided during sudden loss of load.

2.6   Other renewable energy sources

Landfill gas, biomass, municipal waste, etc., are treated as other renewable energy sources for generation of electricity. The location of these generators is determined by the availability of these resources. Major drawbacks of these resources are low-energy density, scarcity of resource and difficulty in storing them in large quantities. Since storage is not cost-effective, these generators are normally of small capacity and operate in load pockets close to the resources. A competitive arrangement,

NFFO (Non-Fossil Fuel Obligation) was created in the UK to encourage generation from renewable energy resources in the late 1990s. The NFFO scheme has been replaced by another support mechanism named as Green Certificates in the year 2000, imposing some obligation on the electricity suppliers' liability to generate a minimum percentage of their total generation from renewable sources. Other countries have also adopted different strategies and sometimes direct government intervention to encourage similar generation schemes.

2.7   Storage devices

The backup energy storage devices that must be included in Microgrids to ensure uninterrupted power supply are

(1) storage batteries,
(2) flywheels,
(3) ultra capacitors.

These devices should be connected to the DC bus of the Microgrid and provided with ride-through capabilities during system changes.

2.8   Conclusion

The basics of some of the DER technologies are briefly discussed in this chapter along with their advantages and disadvantages. The readers may go through to the corresponding bibliography for further details on the technologies. Mostly the popular DERs namely CHP systems, WECS, small-scale hydroelectric generation, solar PV and other renewable energy sources are discussed. Though storage devices are not typical DERs, but for exploiting maximum benefits from Microgrids the application of several storage devices is mandatory with proper demand side management. Some of the research outcomes of the authors are incorporated in the appendices as examples of Microgrid research. The major impacts of Microgrids are discussed in the next chapter.

Chapter 3

Impacts of Microgrid

3.1 Introduction

Microgrids appear to the main grid as aggregated units of loads and microsources. As discussed in Chapter 1, Microgrids are designed as small-scale, low voltage (LV) combined heat and power (CHP) networks for supplying electrical and heat loads to small pockets of customers. More than one Microgrids may also be integrated to form power parks for supplying larger load pockets. The microsources connected to a Microgrid use diverse types of low-carbon generation technologies as discussed in Chapter 2. Microgrids are normally operated in two modes: (i) stand-alone mode as autonomous power islands and (ii) grid-connected mode in synchronism with the main utility grid as per grid rules and regulations. In both modes, they ensure significant benefits to customers as well as to the main grid. Microgrids have enormous impact on main grid operation and its customers. This chapter discusses the technical, economical and environmental impacts of Microgrid. It covers aspects of electricity/heat generation and utilisation, process optimisation, and electricity and gas market reforms to accommodate Microgrids for their potential environmental benefits.

Effective utilisation of waste heat in CHP microsources is one of the potential benefits of a Microgrid. Good co-ordination between heat production, efficient heat utilisation in heat ventilation air conditioning (HVAC) units, chillers, desiccant dehumidifiers, etc., and thermal energy storage (TES) requirements is required for energy optimisation in Microgrids. This can be achieved by incorporating heat generation control and thermal process control features in central controllers (CCs). Similarly, process optimisation functions to enhance overall system efficiency and reliability can also be built in the CCs.

Microgrids have remarkable impact on existing electricity and gas markets. To harness their benefits fully, their market participation must be encouraged. Suitable market reforms must be made to allow such participation, and financial incentives should be provided for owners to invest in Microgrids. Major changes in conventional electricity market have already been initiated in some countries.

Once market participation is assured, there is a vast opportunity for Microgrids to supply quality service to the main utility distribution system. They can naturally provide significant ancillary services to the utility, such as voltage regulation microturbine, a desiccant-type dehumidifier and absorption chiller. The idea is that the microturbine would generate power for the building while the waste heat in the exhaust would supply the absorption chiller and dehumidifier. From utilisation point of view, the main advantage of using CHP sources in air conditioning applications is that the peak demand for air conditioning usually coincides with the utility peak demand for power.

3.2.2   Absorption chillers

The absorption chillers utilise a condenser, an evaporator and a thermal compressor system. It is similar to a conventional vapour compression system, except that it uses a thermal compressor in place of motorised compressor. The thermal compressor consists of an absorber, a generator and a small pump. It takes in low-pressure refrigerant at its suction end and delivers a high-pressure refrigerant at its discharge end. The high-pressure refrigerant vapour (water or ammonia) passes from the generator of the compressor to the condenser. This vapour is then condensed into a liquid and the heat of condensation is released into the atmosphere. The liquid refrigerant passes through an expansion valve that reduces its pressure, consequently decreasing the boiling temperature. The low-pressure refrigerant is then moved into the evaporator, where the liquid boils by drawing heat from the chilled water stream flowing through the evaporator. This in turn cools the chilled water. After this, the low-pressure refrigerant vapour is passed from the evaporator to the absorber. Absorption process takes place at the absorber. A strong absorbent solution with a relatively strong affinity for refrigerant is added to the absorber as a part of the process. Usually lithium bromide/water solution or ammonia/water solution is used for this purpose. During absorption, the low-pressure refrigerant vapour and the strong absorbent gradually combine to give a weak absorbent solution. The weak solution is then sent through a solution pump to the generator where heat is added to boil out the refrigerant from solution. The strong solution is sent back to the absorber after reducing its pressure through an expansion valve and the high-pressure refrigerant vapour is then passed to the condenser to repeat the process.

The operating efficiency of the chiller may be improved by placing a solution heat exchanger between the weak absorbent solution stream entering the generator and the strong absorbent solution stream leaving the generator. This internal heat recovery mechanism reduces the quantity of heat input required to boil the refrigerant out of the solution, thereby increasing its efficiency.

3.2.3   Desiccant dehumidifiers

A desiccant dehumidifier consists of a desiccant wheel filled with desiccant material, and the process and regeneration air circuits. During operation, process air is made to flow through the desiccant wheel for removing the moisture from the air stream before it enters the building. The desiccant material is restored to its dry state by exposing it to the hot regeneration air stream as the desiccant wheel rotates. After that, the regeneration air is discharged to the atmosphere. When moisture is removed from the air, heat is released and the air temperature rises. The hot air is then cooled with heat recovery devices and conventional air conditioner is then used for the final cooling. The final cooling requires much less energy as the air is already dry and free from moisture. In some desiccant systems, a wheel containing desiccant material continuously dries the air and then routes the desiccant material through the waste heat stream to drive out the accumulated moisture.

3.2.4   Thermal energy storage

TES technique is used to store thermal energy to meet HVAC or other process needs. TES is a peak shifting technique that uses various types of storage mediums like chilled water, ice, eutectic salt, concrete or stone. The planning and utilisation of TES through peak shifting may well be implemented with the CCs by programming them to dispatch to the TES according to the next day's cost of power and weather forecast and then its control during the on-peak period.

The traditional method involves using HVAC chillers with a TES tank to shift peak loads. Therefore, when generation falls short or warm weather is predicted for the following day, the CC may decide to use utility or locally generated power at night to charge a TES tank, while during peak times, it may circulate chilled fluid from the storage tank (through a secondary heat exchanger) for meeting the on-peak cooling requirements.

The CC may be programmed according to any or all the three storage sizing strategies that follow:

(1) *Full storage* – This technique shifts all HVAC demand caused by cooling to off-peak hours.
(2) *Demand-limited partial storage* – This method requires real-time control and reduces the peak demand to some pre-determined level according to the demand imposed by the non-cooling loads.
(3) *Load-levelling partial storage* – This method partly supplies the cooling needs for levelling load demand.

3.3   Impact on process optimisation

Microgrid CCs may be effectively employed for optimisation of HVAC systems and manufacturing processes. This section reviews the possible role of CCs in optimising the operation of these systems with respect to power quality, cost of power, transmission congestion cost, basic Microgrid cost, etc.

3.3.1   HVAC system optimisation

CCs can be used for optimising the overall efficiency of the HVAC system and improving its reliability. The computations required for this should consider the energy efficiency value of the major components of the HVAC system such as

chillers, pumping system, evaporators and cooling towers. The efficiency parameters are expressed as the coefficient of performance (COP) given by

$$COP = \frac{\text{Heat or power output}}{\text{Power or equivalent fuel-value input}} \qquad (3.1)$$

The COP depends on several factors like load and humidity. COP values for
various processes of HVAC system are usually as follows:

(1) Efficiency of conversion of purchased natural gas to useful heat $\cong 0.85$.
(2) Efficiency of conversion of residual heat to useful heat $\cong 0.7$.
(3) Efficiency of electricity generation of a reciprocating engine $\cong 0.28$.
(4) Efficiency of absorption chillers to reduce electrical cooling load $\cong 1.0$.
(5) Efficiency of electric compressor-driven air conditioning systems $\cong 0.5$.

The CC should have the knowledge of these COPs for assessing the cost of
various combinations of CHP and non-CHP microsources and absorption chiller
technologies. It should also consider the variation of COPs according to ambient
temperature.

Optimal operation of HVAC systems should consider the fact that this equipment significantly contribute to the peak load. Therefore, for peak shaving by
shifting this load, TES should be made readily adaptable to commercial cooling
systems and their central chilled water plants. However, the operation of the TES
again depends on the type of storage, weather conditions, weather forecast, operating temperature, etc. Thus the optimal use of storage would require long-term
contract arrangements and planning for the operation of CHP microsources. Due to
several factors mentioned in the preceding text, the optimisation process can only
be managed by co-ordinating the operation of the CHP source and the thermal
equipments through a CC.

Studies show that in a typical Microgrid model consisting of a shopping centre,
overall cost savings of about 65% can be achieved by using a combination of CHP
and non-CHP microsources. This is achieved even if (i) the electrical load is about
ten times greater than the heat load, (ii) waste heat is of same type and quality and
(iii) waste heat from the CHP sources is not fully utilised. The overall energy
efficiency of the system can be increased to more than 90% just by matching
thermal and electrical loads exactly to the output of the CHP system. However, for
more realistic CHP systems with diverse types of customers, all the real-time
information related to the type and capacity of the heat loads, their operating
temperatures, flow rates, distances, pressures and efficiency curves should be fed
to the CC. These must be updated from time to time for calculating the optimal
operating condition for the HVAC system and the CHP microsources.

Some studies indicate that with increasing complexity in the control philosophy, the energy management systems of the HVAC systems fail to function properly. Their functions become too intricate to be properly understood by the manual
operators. This has led to the development of more sophisticated control strategies
some of which proved to be quite useful on testing. One such control system is
known as Information Monitoring and Diagnostic System (IMDS). During testing,

IMDS was installed at the test building with data visualisation software. The 57 measured and 28 calculated points and the trended data were carefully examined by the operators to understand how the original controls and systems actually worked. It was found that the IMDS could identify operational problems as well as significant saving opportunities. It could also be used to improve the utilisation of the existing controls. The test indicated the importance of data visualisation software in understanding the operation of complex systems. The test also highlighted the prospects of its application for economically optimising the operation of Micro-grids in co-ordination with utility power distribution and building HVAC systems.

Apart from enhancing the energy efficiency of HVAC systems, sophisticated control strategies can also be used for improving the air quality. Air quality includes air temperature, humidity and CO_2 content. Better air quality is directly related to higher operating costs of the HVAC system because in such cases, the air needs to be cooled first for controlling humidity and more air changes per hour are required for controlling the CO_2 content. The cost increases further when a large number of people are staying in a building leading to higher CO_2 concentration of the indoor air.

It is found that for a large housing complex or a college campus, the energy costs for the HVAC system may be reduced by more than 50% by monitoring the CO_2 level and controlling the air quality. This results in almost 34% reduction in total electric energy costs for the building. Air quality control can be easily implemented through the CC of the Microgrid that supplies the HVAC system.

3.3.2 Power quality

Most of the electronic loads are vulnerable to transients, voltage sags, harmonics, momentary interruptions, etc. These are termed power quality or PQ-sensitive loads. Therefore, both power quality and reliability are of paramount importance. Power quality is a relatively new concept that has gained importance only recently. Voltage sags lasting for only one or two cycles were ignored in the past, but these are now treated as outages. Similarly, momentary interruptions, harmonics and phase imbalances now come under power quality concerns. Though power quality problems have huge economic impact on the industrial processes, but due to diverse industrial applications and several power quality measurements, it is difficult to quantify the exact cost incurred by power quality problems. Some important aspects of power quality problems that are directly related to the financial losses due to outages are as follows:

(1) Magnitude and duration of contingency, e.g. any high-voltage surge.
(2) The nature of process interruption and equipment damage.
(3) Frequency of occurrence of contingency
(4) The time at which the contingency occurs, e.g. whether at peak periods or off-peak periods of production.
(5) Predictability of the customers and the advance notice available to them before the occurrence of any expected contingency.

Some studies are undertaken in the USA by Lawrence Berkeley National Laboratory for studying the impacts of the aforesaid aspects on susceptible office/

requirements. This inability is considered as a violation of contract on the part of the Microgrid. Microgrid should avoid this inability to prevent volatility of the market. The aggregators and marketers normally assist both the system operator and Microgrid owners by combining resources with complementary capabilities. In this matter, Microgrid CC may be programmed to regulate electricity and heat generation according to the price signals from the market.

The market system should try to be more flexible by allowing augmentation of generation/storage and more control for loads. This flexibility would ultimately help the bidders to decide regarding their participation just by studying the market forces from time to time. This process would further help the markets to optimise the power system and the individual bidder's commercial ventures at the same time, minimising central planning and control.

Several commercial softwares are now used by microsource owners to perform detailed economic assessments. These tools help them determine the hourly deployment schedules based on rates or wholesale spot market prices on the basis of real-time price data. Some financial analysis programmes also use detailed modelling of the economic considerations. The economic considerations for this modelling are as follows:

(1) Performance of turbines under site-specific conditions.
(2) Daily energy demand for electrical/thermal load in blocks of time.
(3) Database containing weather data to be used for predicting demand.
(4) Dual-fuel configurations with natural gas and fuel oil.
(5) Evaluation of CO, ammonia and NO_x emissions including the cost involved in controlling the emissions.
(6) The cost of standby charges.
(7) Utilisation of thermal energy.
(8) Avoiding demand charges, thermal energy price and emissions.

3.4.2   Gas market and its difficulties

The increasing tendency to use gas-fired power generation systems is putting tremendous pressure on the natural gas delivery system in many countries. This might raise the natural gas demand by more than 30%. Some important consequences that might result from this dramatic increase are as follows:

(1) The price of natural gas will go up considerably due to increase in demand and shortage of supply.
(2) Increase in demand for gas supply will put excess pressure on existing gas pipeline capacity.
(3) Majority of the gas pipeline network in an area will be consumed for power generation.
(4) Potential loss of compression from pipeline interruption or compressor failure on the pipelines will pose maximum threat to reliability of power supply in a particular region.

(5) Shortage of natural gas might have detrimental effect on the service reliability of gas-fired power generators. This may cause electrical security problem for the system operator.

(6) For vertically integrated power utility, a drop in sales is almost always accompanied by a comparable drop in power cost. However, for a distribution company in the deregulated environment, there is no effective drop in operating costs for a corresponding drop in sales. Moreover, a distribution company has a relatively small equity rate base for which a 5% reduction in sales can actually result in a 50% drop in return on equity.

Therefore, a higher penetration of gas-based generators with no guarantee of supply reliability might pose a threat to the distribution companies under current market scenario.

3.4.3   Necessary market reforms

In the existing market mechanism, the retail customers cannot sense the supply-side cost variations, as these are not reflected in retail price signals. But, the cost information necessary for stimulating demand side responsiveness would not be available unless the CC participates in the wholesale power market. As the market situation is complicated, there are several reasons for which the CC might not be able to properly respond to market conditions. Some of the reasons are as follows:

(1) Real wholesale costs may not be reflected in the retail tariff. Had it been so, Microgrids could be extremely price sensitive to electricity and natural gas costs, and could be tremendously responsive to the market forces.

(2) As the wholesale energy market does not have demand side responsiveness from Microgrid end, the CCs usually provide a highly elastic demand curve. This would be of significant value in a day-ahead market, where the CC could guide the Microgrid to bid in 'capacity' based on the weather and process plans for the next day only. This would thereby permit only one day to plan for generation and consumption.

(3) Load profiles that are used to assign wholesale costs to the energy suppliers fail to account for the actual service costs. A flatter level profile having a considerably lower actual cost is not reflected in the wholesale costs.

(4) The existing wholesale market does not recognise the demand side resources as ancillary service providers. The ancillary service markets are yet to be established where the Microgrids could bid in for their services in such markets.

(5) Transmission constraints are not considered in the pricing of wholesale energy services. Only locational pricing reflects the potential benefits provided by the microsources in eliminating congestion in the existing power supply network.

(6) The markets for ancillary services are still in the budding stage. They are also quite unpredictable due to the lack of well-established equitable market rules. Therefore, the price caps and default service rates must be reformed such that they reveal the value of load profile response at different times and locations and also help to create markets for ancillary services to the existing distribution utility.

3.5 Impact on environment

Micro-CHP systems and other low-carbon generators can effectively reduce emissions and environmental warming. Apart from market sensitivity, this is one of the major criteria to support Microgrid operation. To implement eco-friendly operation, the CC should be programmed to make operational decisions based on the lowest net emission production, considering both displaced emissions and local emission from microsources.

If market-responsive CCs are to include 'minimisation of pollution' as an additional criterion for dispatch decisions, it would make their decision-making algorithms more complex. This complexity can be avoided if a reasonable and fair emission tariff is built into the market system. The tariff would value the electricity supplied from the microsources appropriately after considering the net reduction in emissions. In that case a measure of the net emission reduction is available from the price signal itself.

Emission tariff might be structured as a combined function of time, season and location so that at worst pollution times and locations, the tariff would be most attractive. This would then provide a signal to the CC to operate the microsource optimally for minimising emissions. In this regard, environmental policy initiatives and existing regulatory guidance should also be given due importance.

3.5.1 Minimisation of pollutant deposition

The US Environmental Protection Agency sets the ambient air limits for six air pollutants, viz. (i) nitrogen dioxide (NO_2), (ii) carbon monoxide (CO), (iii) sulphur dioxide (SO_2), (iv) lead (Pb), (v) ozone (O_3) and (vi) particulates. Power stations and highway vehicles are the largest producers of NO_x gases. Large gas turbines and reciprocating engines, operating at high temperature, also result in sufficient NO_x production. On the contrary, microturbines and fuel cells have much lower NO_x emission because of lower combustion temperatures. Thus, their application as microsources would significantly reduce carbon and nitrogen compounds and total hydrocarbons (THC).

Emissions for a microturbine depend on its operating temperature, power output and the control of the combustion process. Emissions can be minimised only through rapid and precise control of the combustion process. Such control is best provided by the microturbine's own control system and should not be exerted through the CCs. The CC may only provide the generation set points for the microturbine considering the emission production versus power level and the displaced emissions for both heat and electric power output. For some applications, however, the CC may monitor the remaining oxygen concentration in the engine exhaust. If this is found to be high, the exhaust may be used either in direct heating or as air pre-heater for downstream burners. Microturbine manufacturers usually exert very strict control in order to minimise NO_x production. Some combustion control methods are as follows:

(1) Wet diluent injection (WDI) method where water or steam is injected into the combustion zone to moderate the temperature. But this method increases CO emissions, reduces efficiency and shortens equipment life.
(2) Catalytic reduction with agents like ammonia in the exhaust. This method is expensive and sometimes forms ammonia sulphate in the exhaust.
(3) Use of catalytic combustors with noble metal catalysts that allow high gas flow rates and very low pressure drop.

Efforts are made to develop environmental–economic dispatch algorithms for microsources using atmospheric emissions of NO_x, SO_2, CO, etc., as weighted functions.

For CHP microsources, usually heat optimisation is the first priority and electricity optimisation is secondary. This means that power production is dependent on customers' heat requirement. For large-scale CHP systems, the operating constraints are as follows:

(1) Heat generated must be equal to the heat demand per hour.
(2) Electricity generated in the process should be used to supply the electrical loads, and any extra power needed must be purchased.
(3) The NO_x, CO_2 and SO_2 emissions must be maintained at specified limits.

Shadow prices might be developed to provide appropriate weighting factors to each of the above-mentioned constraints according to the importance of the variables. Shadow prices are mathematical tools used to quantify the importance of each variable. They might be a function of the present real price, the demand, the time of use and the season. Shadow prices might be established on hour-to-hour basis for each type of pollutants for generating both electrical and heat energy. These would be calculated and used by the CC to arrive at the economically optimal dispatch solution using an iterative method.

Seasonal and area-wise variation of emissions should be given due importance for scheduling and controlling the operation of generators. For example, ozone emission tends to rise in late spring and summer when ambient temperature is high. Due to lengthy reaction times, peak ozone concentrations mostly occur significantly downwind of source of emissions. Moreover, ozone tends to concentrate in densely populated areas at considerable distances downwind from urban areas. Thus, during warmer seasons, it is desirable to minimise NO_x production in or near large urban areas to reduce ozone formation. Microgrids using CHP microturbines can be very effective in accomplishing this. Therefore, attractive rate incentives, based on specific pollutant production, displaced emissions, expected temperature, etc., should be provided for controlling hazardous pollution.

3.6   Impact on distribution system

The most promising aspect of Microgrids is their ability to provide ancillary services to enhance reliability of the existing distribution system. Unlike conventional power stations, Microgrids are located very close to the load pockets. This makes

10 minutes). Frequency responsive and supplemental reserves can maintain the system's generation/load energy balance for up to a maximum period of 30 minutes. After that period the customer must take care of its loads through either their own backup or load shedding.

(2) *Supplemental reserve* – Microgrids can provide this service by making their microsources respond at the system operator's request, within about 10 minutes of contingency. It can maintain the energy balance up to 30 minutes after the contingency until backup supply takes up the loads.

(3) *Backup supply* – Microgrids can provide this supply according to some prior arrangement made by the system operator. The system operator should plan beforehand how to utilise this service for maintaining supply to priority and non-priority loads during primary supply failures with due consideration to priority loads. Sometimes it becomes advantageous for some Microgrid generators to provide backup supply for other loads. The 30-minute warning period is quite sufficient for communicating the need for backup supply to the service provider facility and for responding to the system needs. Market price signal should closely reflect the real-time cost of this service in order to encourage its suppliers to offer the service when needed. Real-time electricity prices are volatile and the costs change dynamically with energy balance between load and generation. Therefore, Microgrids can sell these reserves in open market and make substantial profit out of them particularly during high-price period.

3.6.1.3   Regulation and load following

Integrated Microgrids can efficiently provide the regulation and load-following ancillary services for accommodating temporary load variations.

(1) *Regulation* – Generators are equipped with automatic generation controllers (AGC), which adjust generation to load minute by minute to maintain specified system frequency within the control area. This function is known as 'regulation'. This service can be provided efficiently by the microsources, which are connected to the grid and at the same time located close to the load pockets. This helps to avoid physical and economic transmission limitations in importing power.

(2) *Load following* – Load following is the capability of on-line generation equipment to track customer load variations. The main differences between load following and regulation are as follows:

   (i) Load following takes place over longer periods unlike minute-by-minute load tracking performed by regulation. Hence load following can be provided by many generators.

  (ii) Load-following patterns of individual customers are highly correlated with each other unlike individual regulation patterns.

 (iii) Load-following changes can easily be predicted because of weather dependence of the loads and more or less similar daily load patterns. Alternatively, the customers can also communicate to the control centre regarding any impending change in their load usage pattern. Thus information regarding load-following changes can be effectively gathered by applying short-term forecasting techniques.

Due to the above-mentioned differences, regulation is more expensive a service than load following. Regulation involves higher investment in providing the generators with high speed and easy controllability.

3.6.1.4 Other ancillary services

Other ancillary services obtained from Microgrids are system black start and network stability, which are briefly discussed as follows:

(1) *System black start* – Black start is defined as the capability of a power system to restart its generation after a total system collapse, without importing any external power. This restores at least a major portion of the power system to normal service without any external support. In case of necessity, system operators may have voice communications with trained operators to initiate black start. Stand-alone Microgrids can easily sell power for system black start. Black start units should be located where they are useful and also capable of restarting other generators.

(2) *Network stability* – Microgrids can sell the ancillary service of network stability. Low-frequency oscillations take place in long-distance transmission systems and gradually die down by natural damping if it is not weakened by any loss of generation. If the oscillations are not damped naturally, cascade tripping of generators and hence overloading of transmission lines may occur. Microgrids are capable of sensing the low-frequency oscillations and providing adequate damping. This may be accomplished by making the microsource supply power at $180°$ out of phase from the oscillation. The damping effect would become more prominent if a large number of Microgrids are aggregated.

Dynamic modelling studies carried out by the Western Electricity Co-ordinating Council (WECC) in the USA indicate that large inertia of distributed energy resources (DERs) are sometimes beneficial and sometimes detrimental to transmission system reliability depending upon phase lags. Therefore, suitable controllers should be designed for DERs for eliminating the detrimental effects of DER inertia to enhance overall grid stability.

3.6.2 Distribution system issues of Microgrid

Microgrid is basically an aggregate of DERs. The positive impact of Microgrid on the distribution system is enormous due to its dynamic responses to a wide range of local needs. But utilities, network operators, regulators and other stakeholders are still hesitant to allow autonomous operation of DERs as power islands because of their major impacts on utility operation and protection. The area that is critically affected by DER penetration is protection co-ordination of the utility distribution system. Conventional overcurrent protection is designed for radial distribution systems with unidirectional fault current flow. However, connection of DERs into distribution networks convert the singly fed radial networks into complicated ones with multiple sources. This changes the flow of fault currents from unidirectional to bidirectional. Further, the steady state and dynamic behaviour of the DERs also affect the transmission system operation. Various impacts of DG connection on existing utility network protection are listed as follows:

(1) false tripping of feeders
(2) nuisance tripping of protective devices
(3) blinding of protection
(4) increase or decrease of fault levels with connection and disconnection of DERs
(5) unwanted islanding
(6) prevention of automatic reclosing
(7) out-of-synchronism reclosing

Technical recommendations like G83/1, G59/1, IEEE 1547, CEI 11-20 prescribe that DERs should be automatically disconnected from the medium voltage (MV) and low voltage (LV) utility distribution networks in case of tripping of the circuit breaker (CB) supplying the feeder connected to the DER. This is known as the anti-islanding feature. This is incorporated as a mandatory feature in the inverter interfaces for DERs available in the market. As the DERs are not under direct utility control, use of anti-islanding protection is justified by the operational requirements of the utilities. However, it drastically reduces the benefits of DERs and Microgrids in improving service reliability. Therefore, these issues must be critically assessed and resolved, and market participation of DERs and Microgrids should be allowed to exploit their full benefits.

3.7 Impact on communication standards and protocols

This section discusses the issues related to the development of standards, protocols and communication infrastructure for Microgrid components and controllers. It also discusses the functioning of communication gateway for providing connectivity between different devices and the challenges for developing cost-effective, reliable and standardised communication gateway for Microgrid applications.

3.7.1 Protocols, communication procedures and gateways

Well-structured and universally compatible communication procedures are required for co-ordinating Microgrid operation in stand-alone and grid-connected modes. The procedures should follow the bindings and obligations imposed by the independent system operator (ISO) and power supply authority. In general, computerised control systems need rigidly defined and structured procedures for communication but this might vary from application to application or from authority to authority. That is why the procedures used by ISOs, local generators and heat extraction equipments and local energy management systems are all so different that they cannot provide any connectivity amongst their components. Therefore, a primary issue to be addressed while developing the CC scheme is providing a translation service for different communication methods. It should provide a common basis for communication for all systems.

The device that executes the translations is called a gateway. The main function of the gateway is to provide necessary connectivity amongst devices by message translating, formatting, routing and signalling functions. However, gateways might also introduce unwarranted time delays and other problems in the communication process during information exchange. Some problems are highlighted as follows:

(1) Time delays might be introduced not only by the processing time required for translation, but also because connected networks might have different transmission speeds and different rules for gaining access to the media.
(2) The gateways are designed to pass only a limited amount of specific data. They generally poll devices for this data and then store a local copy. When this data is requested, the gateway prepares its response based on the local copy and not the current data. It may result in supplying old and misleading data, especially for large and complicated systems.
(3) Presence of gateways makes problem solving more difficult because different tools are then required to see and interpret the protocols on either side of the gateway. Moreover, any ambiguity introduced by faulty translation might make troubleshooting even harder. Limited data accessibility through the gateway might also hamper the troubleshooting process, as it may not be possible to access all the data needed to diagnose the problem.

3.7.2   *Alternative communications*

To avoid the problems posed by gateways, ISOs are imposing stringent technical requirements on the gateways used by generators. The California Independent System Operator (CAISO) has imposed different gateway requirements for different types of generation services. For example, the strictest requirements are for the Remote Intelligent Gateway (RIG) used for generators delivering regulation or ancillary services. Requirements are much simpler and lenient for the Data Processing Gateway (DPG) that is employed for generators delivering non-automated dispatch system (non-ADS) ancillary services only. ADS is a gateway for sending dispatch instructions to the market participants on a much slower, e.g. hourly, time frame. For the ADS, usually the ISO provides the software, security cards and card readers while the Microgrid owner provides his own computer and the operating software.

Normally, the cost for a gateway is proportional to the time frame of its operation. Faster gateways used in critical, real-time applications are much costlier than the simpler hourly type gateways. The simpler gateways are just software packages provided by the ISO to be run on a local PC. Technical requirements and the cost of gateways should therefore be duly considered for developing a CC scheme. The main challenges are that the standardised gateway must (i) be reliable and cost-effective, (ii) allow the Microgrid to provide the ancillary services on a faster time frame, (iii) meet the standards for a typical utility supervisory control

Chapter 4

Microgrid and active distribution network management system

4.1 Introduction

Microgrids require wide-range control to ensure system security, optimal operation, emission reduction and seamless transfer from one operating mode to the other without violating system constraints and regulatory requirements. This control is achieved through a central controller (CC) and the dedicated microsource controllers (MCs) connected to the microsources and the storage devices.

The functions of CC and MC and their co-ordination are detailed in Section 1.5. As the name indicates, MCs take care of the local control functions of the microsources. The CC executes the overall control of Microgrid operation and protection through the MCs. Its main function is to maintain power quality and reliability through power-frequency (P-f) control, voltage (Q-V) control and protection coordination. It also executes economic generation scheduling of the microsources and helps to maintain power intake from the main utility grid at mutually agreed contract points. Thus, the CC not only co-ordinates the protection scheme for the entire Microgrid, but also provides the power dispatch and voltage set points for all the MCs to meet the needs of the customers. Thus, the CC ensures energy optimisation for the Microgrid and maintains the specified frequency and voltage profile for the electrical loads. This controller is designed to operate in automatic mode with provision for manual intervention as and when required. It continuously monitors the operation of the MCs through two significant modules, viz. the Energy Manager Module (EMM) and the Protection Co-ordination Module (PCM). The functions of EMM and PCM are explained in Section 1.5. Network management issues of Microgrids and active distribution networks are discussed in this chapter while the protection issues are detailed in Chapter 5.

4.2 Network management needs of Microgrid

Network management needs of a Microgrid are met mainly by EMM with support from PCM and individual MCs. EMM performs microsource generation control, domestic process control (like heat ventilation air conditioning (HVAC)), water heating and chilling optimisation, and energy storage control, maintaining power quality and industrial processes and providing local ancillary services. Microgrids significantly benefit power utilities by peak shaving and providing several ancillary services. However, at the preliminary stage, an EMM should start executing only the basic controls. It can opt for finer and more complicated ones gradually with intelligent electronic devices (IED) and Ethernet communication protocols.

Network management through EMM should focus on the areas discussed in Sections 4.2.1–4.2.5.

4.2.1   Microsource generation control

Microsources use different renewable and low-carbon technologies to generate electricity. Renewable microsources have minimal fuel cost but need to generate at maximum capacity whenever fuel is available. Microsources running on natural gas, hydrogen, etc., should be run when their operation becomes most economic. Combined heat and power (CHP) microsources also generate heat along with electricity. Therefore, for them heat load must also be balanced. Thus, selection of operating period and operating power level of microsources is quite complicated and depends upon cost of fuel, cost of deferral of electric power and heat, and impact of emissions and deferred emissions. Most important goal of microsource generation control is to ensure maximum possible energy saving.

Generation control of CHP microsources must prioritise heat or electric energy because the demands for these loads might not coincide all the time. Sometimes microsources provide ancillary services like voltage regulation, spinning reserve, peak shaving and riding through grid voltage sags. These services bring sufficient economic benefits, and are good enough to override the revenue generated from customers. However, the market structure has a significant role to play in assessing and addressing the competing demands such that it encourages the bidders to participate in the competition. The microsource owners would then execute generation control not only for their own needs but also for exploiting the market opportunities.

4.2.2   Domestic process control

Domestic process control system monitors and controls the heating and air-conditioning equipment of a building, viz. central heating systems, refrigerators, fans, dampers and pumps. These equipment might be supplied from CHP microsources of a Microgrid. The heat loads for a building include hot water supply systems, dryer systems, space heaters, etc. Microsources must optimise their heat generation for these services with due consideration to other demands like providing ancillary services and reducing emissions and fuel cost. Several variables must be optimised for HVAC system to ensure maximum energy efficiency, the set points being dependent on weather parameters and cost of fuel, building architecture and occupancy level. Joint optimisation of heat and electricity production will be performed by the EMM without hampering the operation of the dedicated controllers for chillers, boilers, heat pumps, heat exchangers, dampers, blowers, etc.

Heat generation by conventional boilers and furnaces is more economic during the availability of natural gas at low cost whereas it is more economic to run the

CHP sets at their maximum capacity during peak load hours when electricity cost is high. Thus, EMM should collect real-time data from both electricity and gas markets for fuel price comparison as well as data for short-term weather forecast and accordingly plan the operating schedule for all heat producers (including CHP microsource itself) to achieve process optimisation and system reliability. For achieving correct sequence of operation, the EMM must monitor current system status to send proper command signals to all the equipment.

Ideally, the EMM of a Microgrid should address the following control needs with help from the MCs:

(1) Determination of correct schedule of heat recovery from the sources and their control.
(2) Proper utilisation of waste heat by routing the exhaust and water to the heat exchanger.
(3) Monitoring the exhaust gas inlet temperature to the heat exchanger so that the recovery system might be bypassed if the temperature is too low or too high. Adjustable set points might be used for this purpose.
(4) Monitoring the water temperature at the heat exchanger outlet to guard against overheating and also to provide a signal for the variable water flow control.

4.2.3 Energy storage

To ensure uninterrupted supply to priority loads, EMM needs to control the operation of energy storage devices like battery, flywheels and ultracapacitors. In fact, successful operation of the Microgrid is mostly dependent on proper operation and control of the storage devices during contingencies and disturbances. However, these devices are only used for compensating voltage sags in local buses or as backup power source during actual power outages and not to compensate for hourly energy prices or to level out peak loads. As some microsources have low inertia or ride-through capability, these storage devices also help in supplementing the microsources during low voltage transients on the distribution system, motor starts or other short-term overloads, especially for stand-alone operation.

As most storage devices produce DC voltage, they should be connected to the inverters of the microsources for DC/AC conversion. On the contrary, flywheel generators directly produce AC and hence might directly feed the Microgrid bus. Since storage devices must respond rapidly, they should deploy their own local controllers instead of depending upon EMM commands. Some storage devices like capacitors might store power at high density but are restricted to short-term discharges, whereas some like flywheels suffer from low power density but are capable of discharge for a longer time. Hence, these can be coupled with reciprocating engines for providing energy for extended periods.

4.2.4 Regulation and load shifting

Unlike conventional power utilities, the load profile for Microgrid contains short-term peaks due to the nature of use of these loads. This happens because domestic loads like water heater, oven and heat pump are often run at the same time. This

(3) Storage requirement for fast load tracking
(4) Load sharing through P-f control.

They should ensure that the microsources rapidly pick up their share of load when the Microgrid disconnects itself from the utility. MCs should also enable the seamless transition of the Microgrid from grid-connected to stand-alone mode and vice versa with minimum disturbance to both the systems.

4.3.1.1   Active and reactive power control

The microsources may be (i) DC sources like solar PV, fuel cells and storage battery or (ii) AC sources like microturbines and wind turbines. For the first category, DC power is directly converted into P-f (50/60 Hz) AC while for the second one, the AC output at non-standard frequencies is first rectified to DC and then reconverted into power frequency AC through converters. In both the cases, DC/AC conversion takes place through a voltage source inverter that forms the principal component of the power electronic converter.

Figure 4.1 shows the basic scheme for a typical MC consisting of the microsource and the power electronic converter. The voltage source inverter in the converter system controls both magnitude (V) and phase angle (δ_1) of the output voltage ($V\angle\delta_1$) at converter terminal (Bus-1). The microsource supplies controlled power to the Microgrid bus (Bus-2) at a voltage of $E\angle\delta_2$ through an inductor of reactance X. Normally, $V\angle\delta_1$ leads $E\angle\delta_2$ by the power angle δ, where $\delta = \delta_1 - \delta_2$. The active power flow (P) is controlled by controlling δ, whereas reactive power (Q) is controlled by controlling V. The controls are based on feedback loops of output power P and Microgrid bus voltage magnitude E, which are related as per the following equation:

$$P = \frac{3VE}{2X}\sin\delta \tag{4.1}$$

$$Q = \frac{3VE}{2X}(V - E\cos\delta) \tag{4.2}$$

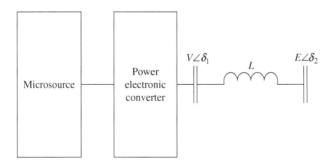

Figure 4.1   Basic scheme for typical MC

4.3.1.2   Voltage control

Apart from active and reactive power control, voltage control at the Microgrid bus is also needed for overall stability and reliability of Microgrids. Microgrids with a large number of microsources may suffer from reactive power oscillations without proper voltage control. Similar to that for large synchronous generators, voltage control function of MC addresses the issue of alleviating large circulating reactive currents amongst microsources. For utility, this circulating current is normally restricted by the large impedance between generators, whereas in case of Microgrids, the problem becomes quite prominent as the feeders are mostly radial with small impedance between the sources. Sometimes, these circulating currents may also exceed the rated currents of the microsources even with small differences in their voltage set points. The circulating currents can be controlled by using voltage–reactive power (*V-Q*) droop controllers with droop characteristics as shown in Figure 4.2.

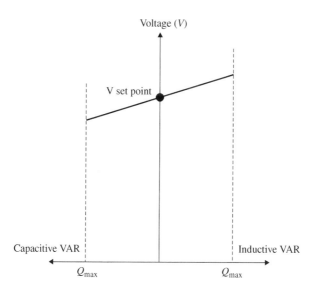

Figure 4.2   Droop characteristics for V-Q droop controllers

The function of the controller is to increase the local voltage set point when the microsource reactive currents become predominantly inductive and to decrease the set point when the current becomes capacitive. The reactive power limits is set by VA rating (VAR; *S*) of the inverter and active power (*P*) output of the microsource as per the following relation:

$$Q_{max} = \sqrt{(S^2 - P^2)} \tag{4.3}$$

4.3.1.3 Storage requirement for fast load tracking

For grid-connected Microgrids, the initial power balance during connection of new loads is taken care of by the large inertia of utility generators. However, for stand-alone operation, the Microgrid needs to ensure initial power balance through its storage devices, which effectively provide the system inertia for the Microgrid. The DC storage devices are connected to the DC bus of the microsource, whereas AC storage devices are connected directly to the Microgrid bus. The MC ensures proper utilisation of the storage devices for fast load tracking.

4.3.1.4 Load sharing through P-f control

Microgrid controllers ensure smooth and automatic change over from grid-connected mode to stand-alone mode and vice versa as per necessity. This is similar to the operation of uninterrupted power supply (UPS) systems. During transition to stand-alone mode, the MC of each microsource exerts local P-f control to change the operating point so as to achieve local power balance at the new loading. The controller does this autonomously after proper load tracking without waiting for any command from the CC or neighbouring MCs. Figure 4.3 shows the drooping P-f characteristic used by the MCs for P-f control.

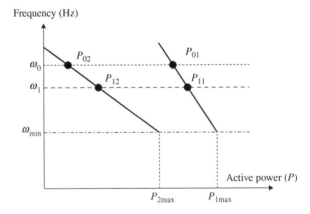

Figure 4.3 Active power versus frequency droop characteristics

During grid-connected mode, Microgrid loads are supplied both from the main utility grid and the microsources, depending on customer needs. When utility supply is interrupted due to any contingency, the Microgrid seamlessly switches over to the stand-alone mode. During change over, the voltage phase angles of the microsources also change, leading to obvious drop in their power output. Hence, local frequency also changes, in which case each microsource quickly picks up its share of load without any new power dispatch schedule from the CC. For example, it is assumed that two microsources operate at a common minimum frequency with their maximum capacities P_{1max} and P_{2max}. In grid-connected mode they operate at a base frequency delivering powers P_{01} and P_{02} respectively. With the change in load demand, the microsources operate at different frequencies causing a change in relative power angles, and the frequency of operation drifts to a lower common value with different proportions of load sharing. This occurs as per the droops of the P-f characteristics as shown in Figure 4.3. Since droop regulation decreases the Microgrid frequency, the MC needs to incorporate a control function to restore the operation to the rated frequency with proper load sharing.

4.4   Central controller

The CC exerts its control through two basic modules, viz. (i) EMM and (ii) PCM.

4.4.1   Energy manager module (EMM)

The EMM incorporates various control functions for controlling the energy optimal operation of the Microgrid. This section discusses a simple EMM that incorporates the basic priority control functions needed for satisfactory functioning of the Microgrid. The number of control functions can always be increased to achieve finer and more sophisticated control, though adding to design complexity.

4.4.1.1   Basic microsource control functions

A simplified EMM provides only the active power and voltage set points for the MCs, while the basic microsource control is executed through the MCs only. Active power dispatch set point is dependent on financial estimation of fuel cost, electric power cost, weather parameters and the anticipated process operation requirements. Voltage set point is kept within a preset band to ensure proper voltage regulation in the Microgrid.

Voltage control
Microgrid loads and their power factors are normally controlled by changing the voltage magnitude and phase angle of the microsources. To avoid extra complexity in EMM control scheme, local voltage and power factor controls of microsources are executed through the MCs and not through the EMM. EMM only provides the voltage set point to the MCs for certain critical buses of the Microgrid. When distribution feeders are not fully loaded in a Microgrid there may be a tendency of voltage rise on the feeders. To arrest the voltage rise, the MCs constantly monitor the local voltage and provide the feedback to EMM. Following this the EMM dispatches the necessary voltage set points to the MCs to carry out the required voltage regulation. The aim of this control strategy is to make the Microgrid appear to the main utility grid as an aggregate of microsources and loads operating as a controlled unit at unity power factor.

4.4.1.6   Optional control functions for intelligent EMM

More sophisticated and intelligent EMMs can be designed in future by incorporating a large number of control functions like industrial process control and intelligent control of the microsources depending on the type of DER. Intelligent EMMs should have extensive information handling capacity, intelligent power electronic interfaces and sufficient communication networks to communicate with the neighbouring equipment. These EMMs should also incorporate control algorithms based on artificial intelligence (AI) techniques. Moreover, they should have remote monitoring and control facilities. Through these they will collect field data over wireless (RF) networks or Ethernet links from several control/monitoring devices via transducers. In future, features like the Internet and GIS compatibility may also be included in the EMMs.

The EMMs must have data/event logging features to allow authorised operators to obtain necessary information about parameters or operating conditions. The EMMs should also provide manual intervention features. The operators should be allowed to program the basic operation and set points for various processes and to enter their own algorithms and models as per customer needs by overriding automatic control functions.

Sophisticated EMMs should be capable of managing information, providing operation guidelines and set points to the system operator and making decisions autonomously to improve overall system performance. Their operation should aim at maximising availability, maintaining high quality service and minimising downtime. They must also monitor equipment degradation and diagnose process problems. With supervisory control and data acquisition (SCADA) systems, they can also supervise and control industrial processes along with Microgrid generation and storage.

Additionally, intelligent EMMs may be used for (i) providing an overview of the process control systems, (ii) focusing on energy consumption and (iii) analysing energy saving opportunities depending on the time of day, process conditions and weather conditions. They may also be employed to monitor the power consumption for electrical loads in a premise and use this data for assessing thermal equipment efficiencies under full and part loads. From economics point of view, they may even be used to automatically optimise the utilisation of microsources and storage by using real-time market price signal for electricity and fuel.

4.4.2   Protection co-ordination module (PCM)

PCM supervises the overall protection for the Microgrid. Protection philosophy for Microgrid is different from conventional distribution networks though both are radial systems. This is because of the following differences:

(1) Microgrids contain both generators and loads resulting in bidirectional power flow through the protective devices in a radial system.
(2) Passive distribution network turns into an active one due to the presence of microsources.

(3) Microgrids undergo a considerable change in its short-circuit capacity when it changes from grid-connected mode to stand-alone mode. This has profound effect on conventional overcurrent relays that operate on short-circuit current sensing.

A key feature of PCM is its ability to distinguish between the protection requirements for the two operating modes and address the contingencies accordingly. Basic protection requirements for the two modes are briefly discussed in the next section with respect to some possible occurrences. Typical Microgrid configuration as shown in Figure 1.1 is considered. However, additional protection features may be incorporated in PCM, depending on the customer-specific requirements.

4.4.2.1   Protection scheme for grid-connected mode

In grid-connected mode, the PCM detects and acts for five possible events. These are (i) normal condition, (ii) Microgrid feeder fault, (iii) utility fault, (iv) Microgrid bus fault and (v) re-synchronisation. PCM takes into account the response times of individual microsources, as well as that of the PCC (point of common coupling) circuit breaker CB4.

(i)   Normal condition
Under normal condition the Microgrid remains connected to the utility through the PCC circuit breaker CB4. Breakers CB1, CB2 and CB3 connect feeders A, B and C respectively to the Microgrid bus. All the breakers remain closed during normal grid-connected operation. The loads are jointly fed by the microsources and the utility.

(ii)   Microgrid feeder fault
In passive radial distribution networks, fault power flow occurs in one direction only, from source to the fault point. So feeder faults are simply cleared by opening the feeder breaker. Fault sensing is done based on fault current magnitude only and not its direction. But as Microgrid feeders contain generators, bi-directional power flow occurs into a feeder fault from all microsources on either side of the fault point. If such fault is not cleared in time, all the microsources may be disconnected from that feeder by their own MCs leading to an extensive loss of generation. To avoid this, feeders A and C are sectionalised into zones by sectionalising breakers. The breakers contain directional overcurrent relays to detect the faulty zone and clear the fault. If the faulty zone contains any microsource, then it is disconnected from that zone by its own MC but continues to supply its local loads connected to the microsource bus. For this operation, the PCM grades the relay settings for all the relays such that the faulty zone is isolated before all the microsources are disconnected from the feeder or before the entire Microgrid gets disconnected from the utility. This strategy ensures minimum loss of generation and Microgrid stability. However, for faults in Feeders A or C, if all the microsources connected to the feeder are located on one side of the

fault, then all will be disconnected by their MCs and the faulty feeder will be disconnected from the Microgrid bus by opening its circuit breaker. If fault occurs in Feeder B, then it can be simply disconnected by opening breaker CB2, since Feeder B does not have any microsource.

(iii)   Utility fault

For any utility fault, the Microgrid disconnects itself from the utility grid by opening CB4. The protection strategy for this case is quite simple. CB4 relay monitors the current magnitude and direction on each phase and sends a trip signal to CB4 if current limits are exceeded within a preset time. Relay setting is provided by the PCM to ensure that the isolation is accomplished without any significant interruption to priority loads. This scheme also ensures that the microsources are not spuriously tripped before the Microgrid separates itself from the utility. Otherwise, this may cause unnecessary loss of generation and reduction in service life of breakers.

(iv)   Microgrid bus fault

If fault occurs on the Microgrid bus, then the Microgrid is disconnected from the utility by opening CB4. Also the Feeders A and C are disconnected from the bus by opening CB1 and CB3 respectively. In case of any fault within the Microgrid, the PCM grades the CB4 relay to co-ordinate with the 'upstream' protection in the utility. CB4 relay is also graded with respect to the protective devices for the microsources to minimise loss of generation, supply interruption and spurious tripping.

(v)   Re-synchronisation

When normal service is restored in the utility, then the PCM's responsibility is to synchronise and reconnect the Microgrid to the utility through synchronism check schemes. This is accomplished as soon as the grid stabilises and goes back to the normal operating state after picking up all previously disconnected loads. However, this may require several seconds to minutes, depending upon the nature of the feeder and loads. The PCM includes a control scheme to bring all microsources into synchronisation with the utility by measuring the phase voltage magnitudes and phase angles, frequency and phase sequence on both sides of the breaker CB4. The PCM provides both the options for automatic and manual re-synchronisation as per necessity.

4.4.2.2   Protection scheme for stand-alone mode

When the Microgrid operates in stand-alone mode, the short-circuit level at the Microgrid bus reduces remarkably. This is because the microsources with power electronic converter systems may only supply up to 200% of the load current to a fault. Thus, a stand-alone Microgrid provides much lower fault currents, as compared to a grid-connected Microgrid. Low fault currents may not be adequately picked up by standard overcurrent relays used in conventional protection systems. Therefore, this is likely to have significant impact on the fault detection capability of the protective relays of the Microgrid based on fault current sensing. Standard relays may take a long time to pick up lower fault currents or may not respond at all.

Thus, alternate fault detection schemes like impedance protection, differential current/voltage relaying, zero sequence current/voltage relaying, directional overcurrent/earth-fault schemes may be adopted for protection of stand-alone Microgrids. Basic operation philosophy for the PCM is briefly described in the following text with respect to the possible events.

(1)   Normal operation
Under normal operation, the Microgrid operates autonomously. PCC breaker CB4 is open. Feeders A, B and C remain connected to the Microgrid bus and their loads are fed from the microsources.

(2)   Microgrid feeder fault
The feeder protection for stand-alone Microgrid is similar to that for grid-connected Microgrid. The only difference is that the relays should be more sensitive to detect lower fault currents. The main objectives of protection are minimum loss of generation and supply interruption.

The design and operation of PCM must take into account the complexity of the Microgrid scheme, the number of microsources, the types of generation technologies, the number of priority loads and the response characteristics of the protective devices.

4.4.3   Information needed for central controller operation

Microgrid operation is dependent on a set information, viz. tariff and regulatory details, microsource performance and diagnostics, weather and load forecasts and heterogeneous nature of service. Some data can be easily captured economically whereas some are quite expensive to capture. In order to optimise the cost of data capturing, the information must be graded according to the relative importance of data in Microgrid control. Due consideration should also be given to the feasibility, procurement and costs of acquisition and processing of the data and the possibility of acquiring alternative data through estimation techniques. Some data important for CC functioning are discussed in Sections 4.4.3.1–4.4.3.5 with regard to their usefulness and cost effectiveness.

4.4.3.1   Tariff, price and regulatory information
For minimising total energy cost for Microgrid operation, cost of energy from utility and from microsources must be compared. The comparison data would be helpful in designing and implementing a suitable energy optimisation strategy through CC. Information on costs of electricity from utility, natural gas, propane, heating oil and biofuels is necessary for CC operation. Electricity purchase from utility is subject to different tariff structures; e.g. traditional static utility tariff with or without (i) energy usage time and (ii) demand charges, and dynamic utility tariff as per real-time pricing or day-ahead market or energy imbalance market as practised by ISO. Microgrids may economically benefit from the appropriate use of interruptible tariffs, emergency demand response incentives or other schemes that would lower their demand firmness. This would however make the economic dispatch strategy for microsources more complicated due to the unavailability of some

(iii)   Reschedulable demand

The load demand that permits staggered rescheduling is known as reschedulable demand. Load rescheduling helps to flatten the demand curve by replacing one single high-power peak with several moderate-power peaks. This is done by shifting energy-intensive activities to off-peak periods.

Through load rescheduling, Microgrids can alleviate microsource overloading and avoid higher demand charges. When the EMM is scheduling the future rather than real-time loads, then the loads must be rescheduled backwards in time. For example, the load of an air-conditioned building may be rescheduled for pre-cooling in off-peak period rather than waiting until cooling demand starts at energy-intensive period. Extent of rescheduling any load is specified by load demand, maximum span of acceptable rescheduling time, cost of rescheduling, allowable time span before start of rescheduling, rate at which load goes down during rescheduling, etc. However, such demand control leads to substantial complexity in EMM design and operation. To implement this, loads should have intelligent controllers for curtailing or rescheduling and they must communicate their status to the EMM.

4.4.4.   Control strategies for central controller design

CC design and operation needs a control strategy to meet the management needs of the Microgrid. The choice of control strategy for CC depends on the performance efficiency of the strategy and cost of its implementation. Some suitable control strategies are real-time optimisation, expert system control and decentralised control, which are briefly discussed in the following sections.

4.4.4.1   Real-time optimisation

Real-time optimisation is the most suitable control strategy for EMM design. Constrained optimisation problems are extensively used in various fields of operations research. In an optimisation problem, the system to be optimised is expressed mathematically as an objective function that has to be maximised or minimised, subject to some constraints. Standard algorithms are available for solving optimisation problems. These algorithms employ intelligent search technique to reach the optimal solution. Instead of evaluating the objective function at every allowable state, the search techniques use their knowledge of system structure to consider only a small fraction of allowable states while determining the optimal state for the system.

Optimisation problems can be formulated as linear programming (LP) or non-linear programming (NLP) problems. The LP problems are the fastest to solve and guarantee a solution. The objective function and constraints for an LP problem are all linear equations and therefore reduces computational burden on the system. However, problems with integer constraints are difficult to deal with and may not guarantee a practical solution. Integer constraints involve yes/no decisions and quantities like the number of machines and/or people, which are only whole numbers. NLP problems are much more complicated and add to the complexity of the controller. NLP algorithms are applied for systems with quadratic objective functions with linear constraints, separable systems or convex systems.

For optimisation strategy, EMM should collect information regarding Microgrid variables such as operating states (voltage/current/power levels), load demands, quantity of electricity and heat generation, open/close status of the circuit breakers, feeder loadings, weather parameters, tariff and operating state/condition of the equipment (normal running/shutdown for maintenance/alert). The EMM must initially consider all past and present variables, stochastic description of loads, weather, tariffs and equipment in order to predict future operation states for the Microgrid. After that it would dispatch a decision to the Microgrid equipment. For simplification of control algorithm, some information may even be approximated by constant values, provided such approximation does deviate too much from actual behaviour.

Mathematical description of Microgrid as a real-time optimisation problem is a complicated affair. This is due mainly to the non-linearities in part load equipment performance and tariff structures (such as demand charges), integer decisions in operating/shutting down microsources, uncertainty about the future load profile, rescheduling options, equipment start-up times and costs, and equipment ramp-rates. Therefore, depending on accuracy of solution required, computing power of the controller and time constraints, the EMM should use approximations to reduce complexity in modelling. It may also use its own optimisation algorithms or commercial software to select the optimal dispatch state and scheduling details.

4.4.4.2   Expert systems

Real-time optimisation puts too much computational burden on the CC and takes a long time to reach a solution depending on the complexity and non-linearity of the system. The cost of implementation in real-time environment is also very high. To alleviate these, AI techniques like fuzzy logic may be used for CC design. AI systems simulate human reasoning and the control algorithm is programmed with a series of decision-making 'If-Then' statements. An expert controller goes through a finite set of control options and simply takes a decision according to some rule bases. Fuzzy logic systems are variants of conventional expert systems. They consider 'overlapping' categories of states and assess 'to what extent' the system belongs to a particular category. Here, several 'If' conditions are satisfied simultaneously but to different extents and the final control decision is a weighted function of the respective 'Then' statements. Fuzzy controllers allow more complexity in their rule bases than conventional expert system controllers.

For Microgrid applications, the functions of a fuzzy-based EMM would be (i) to assess the state of the Microgrid, (ii) to determine what predefined category the current state falls into and (iii) to follow and implement the dispatch rule associated with that category. For optimisation, states should be categorised and the rules specified for them before EMM implementation. After that, adaptive control strategies may be employed for redefining the rules to suit system management requirements.

A major benefit of fuzzy control system is that it resembles human reasoning. Therefore, even if it is not possible to categorise the states and specify rules for the categories at the beginning, it would be possible to develop the controller just by applying the logic followed by human operators in making optimal dispatch decisions.

4.4.4.3   Decentralised and hierarchical control

Decentralised and hierarchical control strategies can be used for (i) aggregating individual Microgrids to bid their excess capacity to the utility and (ii) aggregating individual microsources to bid their capacity to the customers in the Microgrid. In this control, decision-making follows a hierarchical structure. A single agent collects demand and supply bids from multiple agents and makes dispatch decisions for individual agents according to preset rules. The rules for the agents may be determined by a higher-level controller. In case of Microgrids, the individual energy customers and suppliers would act as the agents and report to the ISO. The ISO would then determine the dispatch according to these reports and regulations provided by governing authorities.

4.5   Conclusion

There are some key issues that require extensive research to improvise the design of a Microgrid management system and to make it intelligent in the true sense of the word. Major issues like market reforms, impacts on distribution system, emission reduction, communication infrastructure needs, ancillary services and protection co-ordination have been discussed in detail in Chapter 3. This chapter details how and to what extent these may be taken care of by the Microgrid controllers. Chapter 5 discusses in detail the protection systems in Microgrid.

Chapter 5
Protection issues for Microgrids

5.1 Introduction

A Microgrid is an aggregation of electrical/heat loads and small capacity on-site microsources operating as a single-controllable unit at the distribution voltage level. Conceptually, Microgrids should not be thought of as conventional distribution networks with additional local generation. In a Microgrid the microsources have sufficient capacity to supply all the local loads. Microgrids can operate both in synchronism with the utility (grid-connected mode) and in autonomous power islands (stand-alone mode). The operating philosophy is that under normal condition the Microgrid would operate in the grid-connected mode but in case of any disturbance in the utility, it would seamlessly disconnect from the utility at the point of common coupling (PCC) and continue to operate as an island. Figure 5.1 shows the protection system for a typical Microgrid network.

This chapter reviews two major protection issues that must be dealt with to ensure stable operation of a Microgrid during any contingency. These are as follows:

(1) To determine at what instant the Microgrid should be islanded under a specific contingency.
(2) To sectionalise the stand-alone Microgrid and provide the sections with sufficiently co-ordinated fault protection.

Although the characteristics and performance of most protection elements of a Microgrid are consistent with those present in the utility distribution systems, it is not the same for the microsource power electronic inverter systems because of the following reasons:

(1) Characteristics of inverters may not be consistent with the existing conventional protection equipment.
(2) Different inverter designs have different constants and therefore, do not have any uniform characteristic that would represent inverters as a class of equipment.
(3) Basic characteristics of the inverter unit as seen by the system can change markedly depending on design and application.

portion of its load. If islanding does not occur at the PCC, then the Microgrid may carry a part of the utility with it. In this matter IEEE standards suggest minimum interconnection protection criteria, which a Microgrid should meet during grid-connected operation. Cost of implementation and technical limitations are important for designing protection schemes for Microgrids. Following issues should be duly considered for islanding of the Microgrid:

(1) Whether speed of operation of protection system need to approach SEMI F47 specifications.
(2) How to minimise spurious separations?
(3) Whether non-fault separations would be allowed for under-voltage, open phase and voltage unbalance conditions.
(4) Separation protection limitations imposed by exporting Microgrids.
(5) Whether re-synchronisation to utility would be automatic or manual particularly in relation to frequency and voltage matching.

5.2.1   Different islanding scenarios

This section discusses the following scenarios related to the islanding of the Microgrid:

(1) Fast separation from the faulted feeder
(2) Prevention of spurious separation
(3) Non-fault separation
(4) Separation from exporting Microgrids
(5) Re-synchronisation.

5.2.1.1   Fast separation from a faulted feeder

One major service provided by Microgrid is uninterrupted power supply to priority loads during any outage. If the loads of the Microgrid are so voltage-sensitive as to require separation times of less than 50 ms (as per SEMI F47 specifications), then it will not be possible for the existing protective equipment to act that fast to clear the fault under any condition. Usually, secure relay time to detect an under- or over-voltage is up to two cycles and a medium voltage (MV) breaker requires three to five cycles to interrupt the circuit, after receiving the trip signal. Therefore, if the Microgrid does not have a very fast acting solid-state circuit breaker at PCC, other means must be adopted to prevent the voltage from falling below 50% for three cycles or longer. To achieve design and protection improvements, following two cases have been considered:

(1) When separation is not necessary
(2) When separation is mandatory.

(1)   When separation is not necessary
Such cases occur when the fault is not located between the PCC and the utility sub-station breaker. For example, a fault causing sag on a sub-station bus may occur on an adjacent feeder fed from the same sub-station. In such cases, one option

of preventing sags is to install electronic sag correctors or replacing the Y–Y connected transformer at PCC with Δ–Y connected transformer and adding a high voltage side breaker. For single phase-to-ground faults in the utility, Δ–Y transformer would ensure that the phase-to-ground voltage in the Microgrid does not drop below 58%. These two options demonstrate how protection considerations and design options must be considered together in developing economic Microgrids.

Installation of electronic sag protectors is costlier. Two types of electronic sag protectors are available, one suitable for shorter protection periods and the other for longer protection periods. The short-term sag protector does not use any energy storage and is typically effective for only about two cycles. On the other hand, the long-term sag protector incorporates an energy storage device. However, upcoming sag protectors do not require significant storage if the under-voltage condition does not last too long or is not less than 50%. Even for zero voltage conditions, most sag protectors can hold up the voltage for three cycles. Therefore, if instantaneous relays and three cycle breakers are installed in all feeders adjacent to the main incoming feeder of the Microgrid, then for adjacent feeder faults, the combination of electronic sag correctors and high-speed relaying should satisfy the SEMI F47 requirements for rapid fault clearance. For faults occurring within the Microgrid, the solution will considerably depend on the existing utility practices such as fuse saving with quick blow fuses. In such cases also the electronic sag correctors should still be able to meet SEMI F47 requirements. If, however, the utility employs instantaneous overcurrent tripping of the feeder for fuse saving purposes, then separation of the Microgrid will be required.

Replacing the Y–Y interconnecting transformer with a Δ–Y one is a cheaper solution, though less effective. Since SEMI F47 allows voltage sags below 70% for only 0.2 seconds, the protective action of the utility must be fast enough to meet the specified fault clearing time. Moreover, Δ–Y interconnection is effective for only single phase-to-ground faults on the Δ side and for loads connected between line and neutral on the Y side.

(2) When separation is mandatory

When a fault occurs on its main incoming feeder at the PCC, the Microgrid must separate itself from the utility. The fault point is 'upstream' to the PCC breaker CB4 as indicated in Figure 5.1. This fault requires high-speed separation under all technical standards (like SEMI F47) and utility protection requirements, without maintaining even a low under-voltage tie to the utility. Thus, extensive efforts are needed to develop high-speed protective devices that would meet SEMI F47 requirements even without long-term electronic sag protectors. Storage requirement for the electronic sag protectors is reasonably reduced if high-speed tripping is employed in Microgrids.

5.2.1.2 Spurious separations

From operation point of view, maintaining a tie between the Microgrid and the utility is highly desirable. But if fault occurs on the tie, then the Microgrid should be separated from the utility using fast tripping devices as per SEMI F47 requirements. Inexpensive protective devices are not secure and may cause false trips and spurious separations. Although technical standards specify mandatory voltage and frequency trip settings for measurements made at the PCC, but these are not good discriminators in pinpointing the exact fault location (i.e. whether the fault is on the incoming feeder to the Microgrid or within the Microgrid itself). Currently, the only reliable method of fast tripping the PCC breaker is to have a transfer trip from the utility grid sub-station breaker.

False tripping problems may arise not only from electromechanical relays and breakers but also from the sophisticated microprocessor-based protection packages acting solely on information available at the PCC. Such packages cannot always determine the fault location in spite of the extreme difference in energy capacity between the utility and the Microgrid. The impacts of false trips on a Microgrid are different from those on a single microsource connected to the utility. For the microsource, a false trip just amounts to losing kilowatt-hour sales for a brief period and the cost incurred due to restart and re-synchronising. On the contrary, for a Microgrid, a false trip means significant exposure to power quality problems. Therefore, the cost of interconnection protection must be examined and justified as an insurance against the loss of potential manufacturing production and not just the loss of kilowatt-hour.

If a Microgrid supplies backup power to its own loads, then spurious separations can be tolerated to a certain extent. Rapid separation from a faulty utility safeguards the Microgrid from getting affected by upstream disturbance and allows it to continue operating unfazed. A spurious separation has little effect on the operations of the Microgrid and the utility as long as the Microgrid is able to restore normal operation after separation. The main advantage of tolerating false tripping is that the relay settings for separation can be defined by voltage and frequency deviations and time durations only, even if these are not very good indicators of the exact location of utility fault. If unacceptable voltage deviation persists longer than the allowable duration, then separation should take place, even if it leads to the outage of the main utility feeder. Thus, relaying problem becomes much simpler than trying to assess the fault location from PCC voltage and current measurements.

However, the Microgrid designer may raise the following points as an argument against such an oversimplified approach:

(1) If Microgrids are to shed non-priority load upon islanding, then allowing spurious separation may cause unwarranted outages to these loads.
(2) For exporting Microgrids, spurious separations would lead to loss of revenue and a period of over frequency operation while the Microgrid frequency stabilises. Moreover, the utility may feel that such frequently interrupted generation would not be very worthy.

5.2.1.3   Separation in non-fault conditions

Low voltages (LVs) may also occur under non-fault conditions. Therefore, whether an LV condition is associated with a fault between the PCC and the utility sub-station may be difficult to ascertain without a high-speed communication between the Microgrid and the utility controllers. For under-voltage, it is generally desirable for the Microgrid and the utility to stay connected, while the latter tries to eliminate the LV, provided it is not caused by any fault that requires tripping at the PCC. The Microgrid and the utility may also negotiate a trip control to co-ordinate with the voltage limits for balanced voltage conditions as specified by the SEMI F47. The trip control may be achieved through communication with either (i) the utility or (ii) the trip restraint system using balanced voltage blocking of single phase under-voltage relays at the PCC when the desired voltage trip levels are lesser than the delayed trip settings as specified by IEEE P1547 standards. However, P1547 currently considers trip levels for 'unintentional islanding' conditions and does not cover 'intentional islanding'. Although such trip restraint can be designed using the currently available devices, but still the under-voltage trip setting would be finally determined by operational restraints of Microgrids, viz. voltage sensitivity of loads and the ability of the Microgrid to recover from the LV condition after tripping.

Some degree of voltage unbalance always exists on distribution feeders even under normal conditions. The under-voltage tolerance limit for the Microgrid (i.e. voltage setting under which the Microgrid will island itself) depends on factors like transformer connections and grounding points within the Microgrid. The sensitivity to voltage unbalance of loads, microsources and other distribution equipment should be considered to determine the criteria for establishing the under-voltage tolerance limit. For voltage unbalances, however, it becomes difficult to ascertain whether their cause lies within the Microgrid or is external to it. This job is further complicated by the ratio of the power supplied internally to the Microgrid to its own load demand. Therefore, an intelligent controller function should be incorporated in the Protection Co-ordination Module (PCM) at the PCC to make the appropriate decision of whether to separate or not to be based on voltage unbalance.

Open phases are generally associated with systems where fuses are located between the utility sub-station and the PCC. Open phases may also occur without any fault, though such occurrences are rare. Detection of non-fault open phases is difficult. Complexity depends on the number and type of transformers between the open phase and the PCC where a three-phase switching device usually exists. Since an open phase causes phase-to-phase voltages to remain at or above 50%, the Microgrid may not be able to detect it as an abnormal condition. However, open phase conditions are considered to be potential hazards to transformers and microsources, as excessive overcurrents/over-voltages may exist across the open phases if not properly isolated with three-phase switching. It may also cause safety problems to utility line workers. For this, allowing a Microgrid to connect fuses between the PCC and the utility sub-station circuit breaker should be either strongly discouraged or even if one is used its design must incorporate protection against open phase conditions.

5.2.1.4 Separation of exporting Microgrids

Exporting Microgrids are not in a position to use simple reverse power relays to determine utility contingency conditions. Also, simple over-/under-voltage relaying schemes may not ensure tripping for utility faults as the exporting Microgrid itself

the microsources can isolate themselves from the MV system and continue to supply their local loads. Rather, for Microgrids with microsources at more than one location, as shown in Figure 5.1, options (ii) and (iii) would be the appropriate choices.

If the relaying at the main incoming feeder at PCC is to have enough time to separate the Microgrid from the utility, then microsource protection system should be adequately time-delayed for faults within the Microgrid to avoid loss of microsource generation due to utility disturbances. If grounding within the Microgrid is sufficient to prevent damaging over-voltages, such time delays may be accepted since the fault current magnitudes are much lower in the stand-alone mode than grid-connected mode. But definitely some feature must be included in the microsource protection systems to act as backup to the separation relaying at PCC.

For complicated Microgrid architecture, protection co-ordination may be best achieved through an intelligent PCM while individual microsource protection can protect the distributed energy resources (DERs) without having to co-ordinate with any other protective device. Such a system calls for line or feeder section protection where co-ordinated protection of the various feeder sections needs to be achieved using relatively simple directional overcurrent relays. But in a stand-alone Microgrid, it is difficult to have location sensitive fault currents of high magnitude. It is also difficult to make the conventional directional devices to automatically adapt from higher fault current levels to lower fault currents after separation. Co-ordination becomes difficult if zero sequence current–voltage product relays are used. Another difficulty is that relatively high impedances of the Microgrid source may cause the operating times of two adjacent relays to nearly match, leading to reduced selectivity. Therefore, proper selectivity has to be achieved by employing differential protection schemes around each circuit segment. Moreover, high-speed communication is required to and from all protective devices that trip any faulted circuit.

5.3.1.2   Low voltage fault clearing requirements

Conventional protective devices operate for maximum fault currents levels of about 2–20 times the maximum load current. They are usually time–current co-ordinated with one another so that the device closest to the fault operates first. The device closest to a fault is known as primary device while the upstream devices away from the fault are known as backup devices. During any fault, the backup devices are adjusted to operate slower than the primary one at the maximum fault current that flows through both of them, even though both detect the fault. Time-graded co-ordination takes the advantage of the natural falling off of fault current as the fault moves away from the source of generation. The falling off depends on the magnitude of impedance of lines/transformers between the generation point and the fault point. For MV and LV distribution networks, inverse-time overcurrent (51) and high-set instantaneous overcurrent (with or without time delay, i.e. 50/2 or 50) relays are used. This makes the fault clearing times more or less proportional to the distance of the fault from the sub-station.

The generation point is termed infinite bus in fault studies. One major problem is that the stand-alone Microgrid is not likely to appear as an infinite bus on the MV side of the MV/LV transformer and the apparent impedance of the Microgrid source may be much greater than that of the transformer. Therefore, the fault current change may be relatively small as the fault moves further into the LV system from the MV zone. Thus, in overcurrent co-ordinated protection schemes as described in the preceding text, transition of a Microgrid from grid-connected to stand-alone mode may slow down fault clearing and limit backup protection. Whether this effect would be significant depends on a number of variables which are as follows:

(1) Whether the time-impedance characteristics of the microsources exhibit sub-transient, transient and/or synchronous time effects.
(2) The value of pickup setting of the high-set instantaneous overcurrent relay with respect to the maximum fault current available from Microgrid.
(3) How much inverse the time–current characteristic of the relays is in the region of fault currents provided by the Microgrid, i.e. how much the operating time changes for a change in the fault current.

Keeping these in view, the PCM either should change/readjust the overcurrent relay settings during switching from grid-connected mode to stand-alone mode or should accommodate the setting requirements for both the modes in a single setting.

Some general observations made regarding the effect of this transition on conventional relaying indicate that co-ordination between a primary and a backup device is not disturbed much by mode transition as the devices are co-ordinated at the maximum fault current flowing simultaneously through both the devices. Moreover, the time margin normally improves at lower fault currents due to inverseness of the time–current characteristics of overcurrent relays. If fault currents greater than 10 times the pickup of the time–overcurrent functions are used, the transition will not have much effect on the timing of these functions as the time–current curves are relatively less inverse at these current levels. However, the time of operation of extremely inverse devices will show a more pronounced change due to the transition. The high-set instantaneous overcurrent devices are most likely to be affected due to the drastic reduction in fault current magnitude caused by the transition. There may even be chances of fault currents dropping below the high instantaneous setting of such devices. In that case, faults that were previously cleared within a few cycles may require a much longer time or may not be cleared at all.

One way of matching the fault current levels for the two modes may be to install a fault limiting circuit breaker at the PCC. But, such system calls for a fundamental change in the protection philosophy though it might make the transition from grid-connected to stand-alone mode of operation more seamless. At present, extensive research is going on for developing economic adaptive relaying schemes for distributed generation applications to protect DERs in both grid-connected and stand-alone modes.

5.3.1.3 Presence of dispersed DERs in a Microgrid

The impact of having dispersed DERs on Microgrid protection system must be compared with that of having a central Microgrid generation facility. These must be taken into account while designing a reliable and secure protection system. If the Microgrid has dispersed generation arranged in the form of a quasi network (i.e. if there is no central generation located at the Microgrid bus at PCC), the changes in fault protection scheme must consider not only the reduction in fault current (typical of a stand-alone Microgrid) but also the chances of bidirectional fault current flow in some feeders. Thus, if there is a fault between a slave (controlled) DER and the master (controller) DER, then the slave must disconnect from the system as per P1547 specification. This separation may be difficult if the protection for the slave DER is already designed to not trip for faults on the utility side of the PCC, because in that case, it would not be able to detect whether the fault is on the utility side of the PCC or the master DER side in the Microgrid. Hence, high-speed communication between the PCM and all Microgrid circuit breakers is the only reliable way to achieve selective tripping.

5.3.2 Protection of microsources

The design of a reliable microsource protection scheme should consider the following issues and extensive dynamic simulation studies should be carried out to address them:

(1) Deciding acceptable voltage and frequency protection tolerances for a stand-alone Microgrid.
(2) Assessing whether there is any need for the anti-islanding protection of DERs and if such protection exists, how it may be disabled or overridden while the Microgrid is operating in the stand-alone mode.
(3) Examining whether the existing anti-islanding techniques may lead to voltage and/or frequency instability if used in a stand-alone Microgrid.
(4) Assessing the needs for an under-frequency load shedding scheme for Microgrid's own reliability and co-ordinating it with the same scheme of the utility.

5.3.2.1 Modification of voltage and frequency windows

Although widening the voltage and frequency windows (i.e. tolerance range) during stand-alone Microgrid operation seems desirable for stand-alone Microgrid with low generation capacity, its impact on the safety of existing equipment must be carefully studied before actually implementing the change. If these windows were originally set as protection boundaries for preventing damage to the connected equipment, then they should better not be changed. But, if they were set as fault and island detection levels, then they may be changed only after extensive study. Moreover, this change should be effected only through the intelligent Microgrid central controller (CC).

5.3.2.2 Anti-islanding

Whether there is any need for the anti-islanding protection for the microsources or whether the anti-islanding controls on their power electronic interfaces should be disabled can only be decided by carrying out extensive dynamic simulation studies. In general, it is desirable to deactivate these controls, unless the ratio of utility generation to microsource generation is too high. However, if these controls are not deactivated the Microgrid might be left with uncontrolled islands. As most anti-islanding controls cause very fast tripping, it might be necessary to have these deactivated instantly on the detection of forming an isolated Microgrid. The most reliable way for achieving this would be to transmit a blocking signal from the Microgrid CC to deactivate the anti-islanding trip signal.

5.3.2.3 Load shedding and demand side management

All power systems are designed to deal with local or system-wide overload conditions caused by contingencies like faults or equipment failures. For handling these conditions, the power utilities normally designate a cluster of loads as non-critical loads and disconnect them to avoid any drop in system voltage and frequency, especially during loss of generation or tie lines. This is done through load shedding and demand side management schemes designed to stabilise system voltage and frequency during disturbances. To have the switching flexibility, the utilities provide the non-critical customers with more favourable rates as incentives. Once an agreement is drawn up between the utility and the customer regarding load shedding, the utilities trip these loads rapidly but selectively through their load shedding systems just to restore supply–demand balance by demand side management.

The loss of generation or ties is usually characterised by system under frequency. In that case, the power utilities employ load shedding schemes with under-frequency relays to shed the non-critical loads so as to restore system frequency with reduced generation. In extreme cases, load shedding schemes based on under-frequency relays and inertia characteristics of the rest of the utility system are also employed to restore normal frequency level. Loads tripped in this manner are general loads supplied from distribution sub-station. The only selection criterion is to avoid supply interruption to critical facilities. Normally, these customers neither get any economic benefit from shedding, nor are notified prior to shedding.

Installing a utility load shedding system in demand side management always has technical, economic and political implications. For a Microgrid, however, the nature of the technical problems depends on whether load is to be shed before or after separation. The economic and political problems might also take new dimensions if load shedding becomes unavoidable in just forming the Microgrid.

If a frequency-shed non-priority load exists in the boundaries of a Microgrid, then its tripping must be co-ordinated with the under-frequency separation point established for forming the Microgrid. This co-ordination again depends on the following:

(1) Whether the stand-alone Microgrid has adequate generation capacity to supply this designated non-priority load.

energised by the microsources. Hence, in that case, grounding and possible over-voltage conditions would be mainly determined by the connection of microsources transformers and by the grounding of the microsources themselves. However, a Y-grounded/Δ-connected transformer (with Δ on the LV side) would keep the MV system effectively grounded with the X_0/X_1 less than 3.0, since the transformer itself provides a ground source. In that case, 80% arresters can be effectively used.

The Y–Y connected transformers are not ground sources in themselves and represent straight through paths for zero sequence current. Therefore, the net grounding condition for the backfeed scenario depends on the grounding of the microsource itself. If it is solidly grounded and zero-sequence impedance exists, then X_0/X_1 ratio would be equal to or less than 3.0, so that the 80% arresters may be safely used. But, if it is ungrounded or impedance grounded, then X_0/X_1 would become very high and the healthy phase voltage on the MV system may even exceed the normal phase-to-phase voltage. In that case, fully rated surge arresters should be used. $\Delta-\Delta$ connected and $\Delta-Y$ connected transformers (with the Y-winding connected to the microsource side) can never be a ground source for the system. For such connections, either a grounding bank should be connected to the feeders for limiting the over-voltages or 80% surge arresters on the entire feeder should be replaced with fully rated ones. Moreover, the effect of the over-voltages on other electrical apparatus in the Microgrid served by phase-to-neutral connected transformers should also be carefully studied.

If the microsource is a synchronous generator, then one way of keeping the feeder effectively grounded (during separation of the microsource for an MV system ground fault) is to ground it through a reactor to keep the X_0/X_1 ratio of the system equal to or less than 3.0. Although it is simpler to solidly ground the generator neutral, but in that case, the magnitude of the fault current for a phase-to-ground fault would be greater than for a three-phase fault as X_0 of the generator is usually less than X_1. Therefore, solid grounding is not permissible for large generators. Small microsources may not be subject to this limitation, but still the manufacturer should be consulted to check whether they could be operated with solid grounding or not.

(2)   Co-ordination of ground relays
Relays, reclosers and fuses connected serially along the distribution circuit should be co-ordinated such that farther a device is from the fault, more will be its tripping time. For a microsource, fault tripping should take place in the following sequence (arranged from lower to higher tripping times):

LV breaker or contactor of the microsource → Fuse of recloser just upstream to the microsource breaker → Main PCC circuit breaker of the Microgrid → Device at the MV side of the distribution transformer → Circuit breaker at utility sub-station

This implies that the operating times for the extreme upstream devices may be quite high for certain faults.

As previously discussed, the protective devices in the Microgrid may not face any co-ordination problem for three-phase, phase-to-phase and solid phase-to-ground faults in grid-connected mode. This is because the ratio of phase fault current to load current would be very high in those cases. But, this ratio may not be sufficiently large for high impedance phase-to-ground faults, which may reduce the sensitivity of fault detection. In this case, the Y–Δ transformers have an advantage over the Y–Y ones, as they are able to isolate the zero sequence circuits. The Y–Δ transformer zero sequence current will not flow in the feeder for faults on the LV microsource system. Thus, the MV ground fault relays may be set with a lower pickup current for faster operation since they need not co-ordinate with the ground relays of the microsources.

(3) Unbalanced feeder loads

Most distribution feeders operate with almost balanced loads under normal conditions. But for some contingencies (e.g. intentional or unintentional opening of single-phase laterals), feeder currents may become unbalanced. In that case, the Y–Δ transformers with LV Δ-winding (at the microsource side) being a ground source is particularly susceptible to the unbalanced load currents. The unbalanced feeder overload condition can be alleviated by installing a reactor in the neutral connection of the Y-winding of the transformer. Insertion of the reactor increases the effective zero sequence reactance of the transformer thereby helping to reduce the percentage of unbalance current flowing through the transformer. The Y–Y transformer is also affected due to feeder unbalances though not as severely as a Y–Δ one.

(4) Ground relaying for feeders

For feeders

Since a Y–Δ transformer with Δ-connected LV is a ground source in itself, it may shunt out some of the zero sequence fault current from the sub-station ground relays. This should not pose a major problem because there would still be enough current to operate the ground relays. But problems may occur if (i) the microsource shorts out enough I_0 to prevent the source ground relay from operating at all and (ii) the sub-station breaker or recloser does not operate correctly for fuse saving. In that case, the reach of the fast tripping element might be pulled back so far that the clearing time of its first trip would exceed the melting time of the overreached fuse.

For microsources

Since Δ–Δ or Y–Δ transformers can effectively isolate the zero sequence circuits of the microsources from the MV system, any zero sequence current or voltage measured on the LV of the transformer indicate a microsource fault. Detection of MV ground fault needs to measure E_0, I_0 or both on the transformer MV side. In that case, for Y–Δ transformer with Y-connected MV, I_0 can be easily obtained by connecting a current transformer (CT) in the transformer neutral. However, for Δ-connected MV, E_0 is to be measured from the open delta of potential transformers (PTs) connected to the MV of the microsource transformer. For Y–Y transformers, MV ground faults can be easily detected by measuring I_0 and E_0 on the LV-side of the microsource transformer. This is definitely a cheaper option as compared to connecting MV devices as required for the Y–Δ transformers. However, it is difficult to differentiate between LV- and MV-side faults just using E_0. Although tripping is required for both the cases, yet confusion may arise in determining the exact fault location.

(5)   Need for grounding transformers

When an effective ground source is to be provided from a microsource interconnected through a Δ–Δ or Δ–Y transformer, installing a grounding transformer
would be a cheaper alternative than changing the connection of the main distribution transformer. In that case, the impedance requirement of the grounding transformer would be largely dictated by the kilovolt-ampere rating of the microsource.
The kilovolt-ampere rating of the grounding transformer may also be much lesser
than the main distribution transformer. Installing a grounding transformer would
also allow to choose an optimum value of X_0, independent of the impedance of the
main transformer.

(6)   Magnitude of LV fault current

The magnitudes of three-phase and phase-to-phase fault currents on the LV system
of a grid-connected Microgrid are not affected by the transformer connection. The
effect, however, is appreciable for phase-to-ground faults. With Y–Y transformer
connection, depending on the type of microsource grounding, the phase-to-ground
fault current is usually between 15 and 25 times the transformer full load current.
For Y–Δ connection this fault current is much lower and is only limited by the
neutral resistance of the microsource. If the microsource is solidly grounded and
supplies a neutral for the LV system, the fault current for the Y–Δ connected
transformer is limited by the impedance of the microsource itself. A Δ–Δ transformer results in the same magnitude of fault current as the Y–Δ bank while the
Δ–Y transformer (with Y-connection on the microsource side) produces almost the
same currents as the Y–Y bank.

5.3.4.2   Choice of grounding system

As there is no preference for any particular transformer connection, the Microgrid
should design its own grounding system to suit its distribution transformer interconnection scheme. If Y-grounded/Δ transformers are used, then the Microgrid
remains effectively grounded even after islanding. But, if Y-grounded/Y-grounded
connection is used, then the effectiveness of grounding will depend on the
grounding systems of the microsources, assuming them to be directly connected
synchronous generators. If the microsources are provided with power electronic
interfaces, then it would be difficult to determine the impedance characteristics for
the Microgrid system under single phase-to-ground fault conditions.

But, in general, the choice of the grounding system will not be dictated by the
MV/LV transformer connection, but also by the ground co-ordination required by
the utility. If required, some means for high-speed insertion of a grounding system
may have to be devised upon separation at the PCC.

5.4   Conclusion

Protection requirements for a Microgrid are quite different from those of conventional distribution systems and conventional DER installations. Unlike them, a
Microgrid has to meet two sets of protection criteria, viz. (i) the interconnection

requirements imposed by the utility or specified by appropriate technical standards and (ii) the requirement of separating from the utility in time to maintain the desired power quality and reliability within the Microgrid. If there is any conflict between these two criteria, then these must be resolved, negotiated or tolerated.

Reliable operation of the Microgrid protection system requires sufficient fault current capacity of the stand-alone Microgrid so that all the overcurrent devices within the Microgrid get a fault current magnitude at least three to five times more than the maximum load current. Fault current to rated transformer current ratios lower than three will also violate the transformer protection requirements of NEC 450. This can be achieved only if the Microgrid contains a large percentage of synchronous generation or inverters with high fault current delivery capability. For too low-fault currents, new protection schemes that are not based on overcurrent sensing must be developed and installed, in spite of some uncertainty regarding their cost effectiveness and efficiency.

High speed of operation of the protective devices is very crucial to reliable operation of the Microgrid protection system. Thus, high-speed communication between the utility sub-station and the main incoming circuit breaker at the PCC of the Microgrid should be established for operation of the previously existing equipment that may not be tolerant to spurious separations. Moreover, a solid-state circuit breaker may have to be installed at the PCC if a separation time less than 50 ms, i.e. three cycles (as per SEMI F47), becomes essential for Microgrid viability. For high-speed tripping, very high-speed relaying (less than half cycle) with very fast tripping vacuum circuit breakers may be another good option if these are available at cheaper prices.

Co-ordination amongst the protective devices at PCC, the utility sub-station and the individual microsources should be achieved through an intelligent PCM included in the CC for maintaining reliable operation of the Microgrid and minimise separation from the utility. This may require developing a distribution system version of the pilot wire line differential scheme used in transmission systems. Development of protection system and the PCM would also require development of software-simulated and laboratory-scale Microgrid models for studying the voltage and current dynamics of the Microgrid before, during and after a fault.

Chapter 6 discusses the development of power electronic interfaces for microsources and their controllers.

is quite comparable to that of the power electronics field. The standardised computer architecture bus interfaces provide the framework for fast development of computer industry. This has inspired the power electronics industry to formulate a standard framework in its design like the design of VLSI and computer architecture. The background and its two subtopics – standard computer architecture and VLSI – are beyond the scope of this book. Nevertheless, in the following two sections standard computer architecture and VLSI have been briefly discussed to visualise the rapid growth of computer architecture in contrast with the slow progress of power electronics industry.

6.2.1 Standard computer architecture

Standard computer architecture consists of a set of buses to interconnect the main processor to other peripheral devices and system components such as memory and I/O devices. It results in higher flexibility and lower cost. A bus is defined as a set of dedicated low impedance lines to communicate control, data and address information enabling straight interconnection between central processor and peripherals. The bus-centred approach is very compact and cost-effective because it connects all subsystems directly to the central processor. The manufacturers of computer peripheral devices can avail the advantage of product economy due to their compatibility with several computer brands by virtue of industry standard bus architecture. Thus, a computer system becomes highly flexible having the choice to select its components from a range of manufacturers, for both new installations and replacements. However, a major drawback of this approach is the bottleneck in data communication through the bus, which is normally taken care of by data multiplexing technology. Besides, this approach forces the customised products out of market leading to another bottleneck in accommodating specific application requirements that cannot be catered by the available industry standard bus architecture.

The bus-centred approach for computer systems and that for power electronic converters are quite different. In computer systems bus characteristics are defined by bandwidth and latency, whereas in power electronic converters it is done by power throughput and power density. Bus scalability is also of paramount importance because of the requirement of gradually increasing power ratings of the converters. Hence, this new venture in power electronic technology must be carefully implemented unlike computer industry for taking care of all the aforesaid factors.

6.2.2 Very large scale integration (VLSI)

In the design of microprocessors, VLSI technology has revolutionised the manufacturing process towards vertical process orientation. The same technology can be applied in the manufacture of electric power converters too. In VLSI technology the designers can describe the system at a high level. Software-based automated engineering process might produce pattern generation files and wafer fabrication layout information leading to foundry process to produce the specified integrated circuits. In this vertical orientation of the design process, the designers can design the system at function level without specifying any process technology. The production of power converter might also be oriented in a similar manner by specifying the design parameters at function level producing a range of standardised components within the rules of architectural framework. The process implementation needs to define the standardised components and connective frameworks in the design and manufacturing environment.

6.3 Power converter trends

The developments in manufacturing process of power electronic converters are lagging behind that of computers and microprocessors. A major challenge for power electronic converters is their high power rating. The types and ratings of power converters are quite large, but the production process is similar. Industry requirement has led to custom design and power modules integration in the manufacture of power converters.

6.3.1 Custom design and manufacturing

The custom design and manufacturing of power converters are application-oriented and tailor-made. Hence, these change for different applications. For accuracy in converter performance, the product design requires circuit simulation and printed circuit board (PCB) layout by suitable software package. Besides, it requires finite element analysis for thermal and electromagnetic behaviour and solid modelling for packaging and assembly. Most of these production layers are normally executed manually in the custom design leading to high cost. The custom design and manufacturing are economically viable only when the converters are relatively of small size and the production volume is very high. The cost of manufacturing setup is compensated when the production volume is very high. In custom design and manufacturing, the replacement of non-functional faulty components becomes too costly an option. Hence, small size converters are practically economical to replace the whole converter unit in case of any component or subsystem failure.

High-rating power converters are used in DG systems as power supply back-up and power conditioning, and in high power electric drives. These are quite large in size and the production volume is very small, making custom design and manufacturing very expensive. Besides, replacement of the whole converter and repair of the converter for any component or subsystem failure are both very costly. Thus, custom design and manufacturing of larger converters are not economically viable.

6.3.2 Power module integration and component packaging

Packaging of power switching devices is an efficient way to improve power converter performance, reliability and power density. Integrated power modules (IPM) incorporate power electronic components like insulated gate bipolar transistors (IGBT) and anti-parallel diodes. These are put in one package with some or all the ancillary power electronic functions, viz. gate drive, protection, auxiliary snubbers, logic, isolated power supply, sensors and digital/analogue control. System components like inductor, capacitor, filters, fans, heat sinks and connectors cannot be integrated into IPM to avoid high assembly cost. Power module integration and component packaging are mainly influenced by application market. This leads to incompatibility and uselessness of readily available factory *de facto* standard packages. Thus, the integration of components to achieve an electrically, thermally and volumetrically compact design is very difficult. It is now realised that co-ordinated packaging efforts are necessary for implementation of IPM, that also partially, for cost-effective low-/medium-power converters.

6.3.3   Power electronic building blocks (PEBB)

Power electronic building blocks (PEBB) approach is used to implement increased power density in multifunctional plug-and-play modules for universal power processors. It proposes modular components of power converters as an extension of IPM development. PEBB approach of component integration is also limited to very selective less-flexible power converter applications. Nevertheless, its co-ordinated approach of integration and virtual test bed (VTB) software engineering environment are good guidelines for BBS framework.

6.3.4   Packaging framework design

Packaging framework design mainly focuses on power electronics packaging, design and development without exploring the prospect of system standardisation. However, it results in several ideas that are useful in BBS framework approach. This is a four-dimensional design approach incorporating user requirements, packaging levels, interfaces and pathways, and four forms of energy. This framework is broadly applicable because it addresses high degree of abstraction, different issues of converter scale and layering, converter elements and connective paths and different forms of energy and energy flow. An object-oriented software for design automation of power electronic devices might be the foundation of a more comprehensive software tool for automated design of integrated modular BBS framework.

6.4   Bricks-Buses-Software (BBS) framework

Power electronic converters basically condition the source power as per load requirements. Previously, in developing power electronic converters, device improvement issues like smaller size and increased switching frequencies were of prime consideration as compared to the ancillary design aspects like connecting pathways, interface mechanisms, packaging and thermal management. Recently, the ancillary design aspects have gained more focus because of their profound influence on designing low-cost high-performance converters.

BBS framework is achieved by abstracting the engineering process to a high level in three components, viz. (i) an array of modular components i.e. bricks, the constitutional blocks of power converter; (ii) a range of connective bus architectures for straightforward interconnection of bricks and (iii) a comprehensive software environment for defining power converter at an abstracted level and translating these information into engineering and manufacturing files as per predefined bricks and buses. BBS framework can be termed power converter design compiler. Electrical, thermal and mechanical interconnections of converter components depend mainly on brick and bus specifications as well as on geometric compatibility in common face alignment. It needs to ensure proper functional operation of each brick and prevent any cross-brick interference, electrical loading, electromagnetic interference, data loss, excessive heating and mechanical misalignment. The aforesaid engineering process executes automated translation of topological converter design into hardware specifications. This can be readily used to manufacture the power converter based on standardised bricks and buses. The BBS framework is helpful in reducing cost, increasing serviceability and bringing about performance improvement and cleaner design and manufacturing cycle.

6.4.1 Bricks as modular components

The BBS framework approach of power converter design depends on commercial availability of bricks as modular components. But its successful implementation is based on the degree to which power converter subsystems can be reengineered and repackaged into bricks. This design facilitates the efficient use of volume for high power and capacitance density as well as straightforward bus connectivity through common face alignments. Typical converter elements like power semiconductor modules in PCB, ancillary gate drive circuitry and gate drive power with attached rectangular air-cooled heat sink can easily be accommodated in designed brick-shaped element. However, some elements must be reengineered to be converted into brick-shaped components. Some useful representative converter bricks are briefly discussed in Sections 6.4.1.1–6.4.1.7.

6.4.1.1 Power switching brick

The power switching brick is the main element of major power converters. It contains power flow controlling semiconductor devices, viz. Thyristors, GTOs, IGBTs and MOSFETs. The shape and packaging of power switching brick depend on the semiconductor technology. A power converter assembly may have different power switching bricks for rectification, inversion and DC voltage conversion. The bricks may also be operated in tandem for achieving necessary power ratings. Besides, they may house some ancillary components like gate drive, power device protection, isolated power supply, signal isolation devices, heat sinks, sensors, voltage decoupling capacitance and snubber networks. IPM or PEBB switching module would be a good foundation for power switching brick for avoiding unnecessary complications in heat dissipation. Electric power connection to power switching brick is made through power bus, whereas thermal bus connection is dependent on the extent of heat generation. Heat generation is again determined by the power level and the characteristics of thermal extraction medium. The design and packaging of power switching brick should avoid any cross coupling and electromagnetic induction by incorporating local voltage decoupling capacitance

6.4.2.2 Thermal bus

Thermal bus extracts heat from hot spots by flow of coolant (air or water) and dissipates it to external sinks. Coolant flow across hot surfaces of bricks is controlled either by embedded control or with control bricks for minimising the effects of thermal loading. In low-power converters, bricks may have independent heat sinks with dedicated controlled fans. However, for high-power converters, thermal buses guide the coolant through network piping to the appropriate brick, thus ensuring reliable and leak-proof heat dissipation.

6.4.2.3 Control bus

Control bus communicates control and sensor data between bricks. Dedicated power lines of control bus carry auxiliary power from auxiliary utility brick. The control bus consists of conductors in the form of ribbon cable, PCB or printed flex circuit and connects bricks via snap-in connectors. It carries 5 V digital logic signals. Control bus also shields its lines from dv/dt and di/dt based electromagnetic induction of power bus. For special requirements fibre optic or wireless connections may be employed. The control bus data format is decided as per control brick convenience with synchronising master clock. CAN or other suitable protocols are used for networked control.

6.4.2.4 Structural bus

Structural bus is a mounting block to house the connections of several buses and bricks of BBS converter. Structural bus design depends mainly on physical converter size. Central beam structure or rectangular structure may be used for housing buses and bricks with standardised aluminium or steel channel pieces.

6.4.3 High-level software design environment

The BBS framework design is usually processed in a high-level computer aided design (CAD) environment. Graphical layout and key design parameters for the basic converter topology are translated into a power converter manufacturing file after providing sufficient information for the automated production of converter by fabrication house. The main design specification includes converter topology, basic control block diagram and design parameters like maximum power, voltage and current, switching frequency, filter bandwidths, voltage and current ripple limits, reliability indices and control objectives.

CAD translation engine produces converter manufacturing data (CMD) file as per basic converter definitions. The CMD file describes in detail the layers of VLSI digital system or PCB. It does not graphically describe a layout defining co-ordinates of a photo etching or exposure tool; rather, it contains a list of bricks used in the converter and necessary information for their connection through different buses. The fabrication house assembles the converter from CMD file using standardised BBS elements by bus fabrication and converter packaging assembly. To integrate any custom component in the design process, a brick definition file is introduced in the design library wherefrom the custom component can be manually inserted into the layout file. Afterwards, the layout translator automatically integrates the manually selected custom component into the generated converter design.

In many cases the design process needs converter performance analysis before fabrication. The CMD file with converter element definitions may be used for verifying converter design at a lower level with the help of electric circuit viewer, thermal circuit viewer, control topology viewer, solid model viewer, etc. If necessary, the low-level converter design data can be extracted for simulation or finite element analysis by any software tool like PSPICE, SABER, EMTP, MATLAB-Simulink, SolidWorks, ProEngineer and I-DEAS to do any modification in the graphical layout for optimal tuning of converter design through iterations.

6.5 BBS framework issues

Several issues should be addressed for realising BBS framework. The approach of packaging elements with connective bus architectures has several benefits and drawbacks with regard to modularity; aspect ratio; parasitic inductance; and cross coupling and loading, including electromagnetic and thermal interference. These are discussed in Sections 6.5.1–6.5.4.

6.5.1 Modularity

Modularity of BBS framework approach lays the foundation of high-level design environment. It allows straightforward integration of converter elements into designs that are compact electrically, thermally and volumetrically. Component packaging standardisation makes component replacement and subsystem upgrading easier. Component packaging flexibility is also suitable for multiple converter applications leading to lower unit costs.

However, unlike custom design the modular approach is not optimum for any specific application. This results in high cost overhead. Medium- and high-power converters can accommodate this cost overhead with large volume of production but not low-power converters. Unlike microprocessor design processes, BBS technical specifications are not continuous in nature because of different converter elements and various manufacturing technologies. This again adds to cost inefficiency.

6.5.2 Aspect ratio

Aspect ratio is the relation between two characteristic lengths of converter element bricks. It is relevant for efficient packaging within a given volume. Efficient packaging of capacitance elements is problematic because capacitance size depends on voltage rating and ripple current to be handled and not much on aspect ratio. The height requirements are conflicting for different bricks, leading to mismatch in aspect ratio. This results in inefficient use of converter volume. Hence, the issue of aspect ratio needs to be addressed for volume efficient packaging of converter in BBS framework.

6.5.3   Parasitic inductance

Parasitic inductance is the unintended product of interaction between bus structure and modular components. It results in performance hindrance of BBS framework converter. The most common effects of parasitic inductance between semiconductor switching devices and decoupling capacitance on a DC bus are voltage spikes, resonance and increased switching losses due to high switching frequency. Bus-centred converter is preferred to custom converter layout as reduced conduction path length between voltage stiffening and power switching bricks results in less parasitic inductances. However, in BBS framework converter there are conflicting requirements for minimising the effects of aspect ratio and parasitic inductance altogether and a compromise needs to be made to minimise the aforesaid effects.

6.5.4   Cross coupling and loading

The bus connective architecture in BBS framework is intended to minimise unintentional coupling and loading between any combination of bricks and buses. The effects of electromagnetic interference and thermal interference due to neighbouring bricks must be minimised.

6.5.4.1   Electromagnetic interference

Electromagnetic interference takes place due to electromagnetic energy radiated from power switching bricks and power bus. The digital signal flow between control bricks and sensor bricks through control bus is affected by electromagnetic interference. Possible remedies may be electromagnetic shielding of sensitive bricks and buses for minimising connective path lengths between power bus, power switching bricks and control bus.

6.5.4.2   Thermal interference

Thermal interference may occur between power switching bricks, voltage stiffening bricks and other temperature-sensitive components of the BBS framework converter. Cooling system should effectively minimise thermal loading in closed loop thermal bus for multiple brick system. Feedback control loop in the closed loop thermal bus may be a possible solution for effective cooling of all the temperature-sensitive elements.

6.6   Conclusion

The BBS framework approach for converter design basically consists of three major elements, viz. (i) bricks – the modular components that are the elements to constitute practical converter topology, (ii) buses – the connective architectures that are the interconnecting links between bricks and (iii) high-level software – environment in which the converter is described at an abstracted level for generating automated engineering and design files.

There are tremendous possibilities of improving the recent converter design with this approach, the advantages being cost reduction, improved serviceability, accelerated performance improvement, cleaner design, manufacturing and assembly cycle.

For further research in this field, some processes must be planned in parallel, viz. (i) development of next generation power electronics design environment, (ii) development of mass customised manufacturing process as per design environment and (iii) development of plug-and-play control approaches on suitable platforms having compatibility with power conversion process.

Mention may be made that power electronics market is quite mature in its own right particularly for motor drives, UPS and several other converters. But the design and manufacturing for most of the existing power electronic devices are not based on modular approach unlike advanced computer and microprocessor architectures. Since the success of Microgrids is very much dependent on the use of efficient cost-effective power electronic interfaces, the modular approach for the manufacture of power electronic devices would be widely accepted with the progress of Microgrids.

Functioning of power electronic interfaces for Microgrids and active distribution networks is directly related to the development of supervisory control and data acquisition (SCADA) and communications infrastructure in the same area. Chapter 7 deals with SCADA and communications in Microgrid management.

SCADA systems are mainly available for controlling networks at voltage levels higher than 6.6 kV with high-bandwidth local area networks (LANs) generally located at remote locations. SCADA functions might consist of data acquisition, data processing, remote control, alarm processing, historical data, graphical human–machine interface (HMI), emergency control switching, load planning tools for demand side management, etc. SCADA systems are quite secure and robust with diverse routes at higher voltage levels, e.g. 132 kV and more. Since DNOs operate in both dense urban and dispersed rural networks, DNO SCADA has to utilise available cost-effective communication channels with diverse characteristics namely analogue circuits, slower bands up to 2,400, duplicate routes for 132 kV key sites and multi-dropped with more than 10 RTUs per line. A combination of several communication circuits is used in SCADA, e.g. private pilot cables, rented fibre and copper circuits, mobile phone technology and radio. The existing communication structure is mainly based on copper cables that are not as reliable as optical fibre ones. But these fibre-optic cables are quite costly and not widely available in rural areas where most of the renewable DERs are based. Thus, communication infrastructure is a major challenge for DNOs for connecting both individual DERs and Microgrids.

DNOs use several RTUs with a variety of functions namely time tagging of switch changes, alarms with a maximum accuracy of 1 ms, digital outputs (DOs) with programmable pulse duration, intelligent relay connections and programmable logical sequences.

7.3   Control of DNO SCADA systems

Since the active network management depends on SCADA systems, the volume of data transmitted over long distances need to be reviewed before implementing the control scheme of SCADA. The increase in length of communication channel might lead to risk on reliable data transmission. Therefore, active network management solutions might be located at strategic sub-stations instead of remotely located DNO control rooms. Normally two control schemes, e.g. centralised control and distributed control, are used. The advantages and disadvantages of these control schemes are discussed in Sections 7.3.1 and 7.3.2.

7.3.1   Centralised SCADA systems

There are some SCADA functions that need to be controlled centrally, e.g. scheduling of load shedding sequences and demand side management. But the major challenge lies with the suitable communication infrastructure for reliable data transmission. The 33/11 V sub-stations might have RTUs in place but the 11/6.6 kV sub-stations might not have RTUs as well as any communication infrastructure. It leads to slow detection of switch changes having sluggish SCADA response. Centralised SCADA systems might have sequence capability, network diagram, asset database, hardware and software maintenance facilities, centrally managed configuration control, etc. that are helpful to implement future intelligent sequential operations. However, it might suffer from the difficulties such as lack of cost-effective communication infrastructure, sluggish response leading to untimely sequential operations, testing bottlenecks for physical distances and risk of single point of failure.

7.3.2   Distributed SCADA systems

The distributed SCADA systems comprise small SCADA systems located at diverse sub-station locations. It could be advantageous with a single facility for management and generation of sequence schemes. These could be distributed through dedicated support and monitoring workstations. However, distributed SCADA systems might have the advantages of cost-effective modular repeatable logics, low-cost communication infrastructure with exception reporting radio system, better response time in switching operations, etc. But it has some difficulties and challenges namely incompatibility with central SCADA system, necessity of additional maintenance facilities, availability of suitable cost-effective management tool for multiple distributed operations, and requirement of field visits for logic modification with additional field staff. Nevertheless, the application is technology dependent and its suitability and cost effectiveness would mainly depend on repeatability and complexity of SCADA solution.

7.4   SCADA in Microgrids

SCADA in Microgrid is a medium-scale, distributed system to monitor and control electric power generation, heat generation, storage devices, distribution and other ancillary services. It consists of input/output (I/O) signal hardware, controllers, networks, communication, database and software coming under the purview of instrumentation engineering. SCADA basically refers to a central control system that monitors and controls a complete site or a system spread out over a long distance. The major control operations are automatically executed by RTUs and/or programmable logic controllers (PLCs). However, the host control functions are usually restricted to basic site override or supervisory level capability. The SCADA system always authorises an expert operator to execute any manual control, if necessary, for overriding automatic control functions. The SCADA system takes in necessary feedback signals from RTU and/or PLC via closed control loop for its monitoring and control purposes. Data acquisition begins at the RTU or PLC level with the communicated signals for meter readings and equipment states as per the needs of the situation. These data are then compiled, processed and made available to a control room operator through the HMI. The operator finally uses the data to make necessary supervisory decisions for adjusting or overriding normal RTU and/or PLC controls. Data can also be collected to build a commodity database management system (DBMS) for trending and other analytical work.

SCADA systems normally implement a distributed database in the form of a tag database that contains data elements known as tags or points. A point represents a single I/O value monitored or controlled by the system. Points are of two types, e.g. 'hard' and 'soft'. A hard point represents an actual input or output connected to the system, while a soft point is the result of logic and mathematical operations applied to other points. In most cases these are treated simply as points. Point values are normally stored as value–timestamp combinations after the value is recorded or calculated. The history of that point is a series of value–timestamp combinations. Sometimes, additional metadata are also stored with tags namely path to field device, PLC register, design time comments and alarming information. The SCADA in Microgrid is basically a distributed control system (DCS) that might be installed from a single supplier, but the common practice is to buy different components from prospective companies and then assemble them through Ethernet communication.

The detailed discussion on the SCADA systems is beyond the scope of this book. However, its basic functions are briefly narrated in the following sections for the understanding of its applicability in Microgrid and active distribution network management with advantages and challenges.

7.5   Human–machine interface (HMI)

HMI is the device that provides the processed data to the human operator for supervisory control actions. The HMI has always been an essential requirement to monitor and control multiple remote controllers namely RTUs, PLCs and other control devices in a standardised way. Normally, PLCs are distributed across a plant to execute programmed, automatic control over a process. PLCs make it difficult for the HMI to collect presentable data directly from them for operator's information. The SCADA system provides processed data to HMI after gathering information from the PLCs as well as other controllers via a standard network. The HMIs are also linked to a database through DBMS for acquiring diagnostic data, scheduled maintenance procedures, logistic information, detailed schematics for a particular sensor or machine as well as expert-system troubleshooting guides. For a long period, major PLC manufacturers have been offering the integrated HMI/SCADA systems that mainly use non-proprietary open communications protocols. Various specialised third-party HMI/SCADA compatible packages are also available in the market for interfacing most of the major PLCs enabling engineers and technicians to configure HMIs themselves without using any tailor-made standard software packages. The main advantage of SCADA is its suitability to a wide range of applications starting from controlling of room temperature to nuclear power plants due to its high compatibility and reliability.

7.6   Hardware components

SCADA system consists of different DCSs namely smart RTUs and PLCs that are quite capable of executing simple autonomous logic processes without intervention of the master computer. The functional block programming language, IEC 61131-3, is widely used for developing programs to run on these RTUs and PLCs. The IEC 61131-3 has minimum training requirements by virtue of its resemblance to historic physical control arrays unlike procedural programming languages namely C and Fortran. It enables SCADA system engineers to perform both the program design and implementation for execution on an RTU or PLC.

The four major components of a SCADA system are (i) RTUs, (ii) PLCs, (iii) master station and HMI Computer(s) and (iv) SCADA communication infrastructure.

7.6.1   Remote terminal unit (RTU)

One of the most important functions of SCADA system is to generate alarms in the form of digital status points indicating the value as either NORMAL or ALARM. An alarm is generated when certain preset conditions are fulfilled. This is generated to draw the attention of the SCADA operator to the part of the system requiring attention for some control action. Backup text messages are also sent along with an alarm activation for alerting managers along with the SCADA operator.

The RTU is a device used for interfacing controlled objects in the physical world for DCS or SCADA system by transmitting telemetry data to the system and/or altering/controlling the objects as per control messages received from the SCADA system or DCS. A typical RTU consists of a communication interface that may be serial, Ethernet, proprietary or any combination, a simple processor, some environmental sensors, some override switches and a device bus or field bus to communicate with devices and/or interface boards. Sometimes the device bus or field bus is used to interconnect RTUs with host systems as well as field devices. The interface boards are capable of handling analogue, digital or both types of I/O signals with different range of inputs, protection capability against voltage surges, intelligence level of the interface, etc. Some RTUs or PLCs are available with integral interfaces to be connected directly to the system without any bus interface in between monitoring and control of a few devices.

The interface boards are normally connected through wires to physical objects under control. In most of the SCADA applications, high-current capacity relays are connected to a DO board for switching field devices. Analogue inputs are usually of 24 V with a current range between 4 and 20 mA. The RTU converts these input data into appropriate calibrated signals to be fed into HMI or MMI (man–machine interface). The RTU uses DO board to execute any control switching operation as per the signals of SCADA or DCS system. Modern RTUs can execute simple programs autonomously without the intervention of the host computers in DCS or SCADA system thus providing necessary redundancy for safety reasons. Modern RTUs can modify their behaviour by toggling to override switches by maintenance operators thus ensuring safety in process operation and control. RTUs and PLCs are gradually overlapping in their operations incorporating similar control features offering proprietary alternatives and associated development environments.

protocols are becoming very compact day by day and many are designed to send information to the master station only through RTU polling. Typical legacy SCADA protocols are mainly vendor specific namely Modbus, RP-570 and Conitel. These communication protocols are all SCADA-vendor specific unlike standard protocols namely IEC 60870-5-101 or 104, Profibus and DNP3. These major communication protocols are basically standardised and mostly recognised by all major SCADA vendors. The extension of most of these protocols to operate through Internet TCP/IP, although avoiding Internet in SCADA communication, is a good security engineering practice for drastic reduction in the attack surface. Since RTUs, PLCs and other automatic controller devices were developed before the advent of industry-wide standards for interoperability, hence a multitude of control protocols were developed for proper communication of SCADA with the available controller devices. In order to lock in the customer base, the larger vendors always try to create their own protocols.

7.7   Communication trends in SCADA

The basic trend for PLC, HMI and SCADA software is more 'mix-and-match'. In the early mid-1990s, the typical data acquisition systems that I/O manufacturers would provide were their own proprietary communication protocols over a suitable-distance carrier namely RS232C and RS485. In the late 1990s, the I/O manufacturers started to offer open communications through support of open message structures like Modicon Modbus over RS485 and by 2000 most of the I/O makers offered complete open interfacing such as Modicon Modbus over TCP/IP. The introduction of Ethernet TCP/IP in industrial automation, e.g. determinism, synchronisation, protocol selection and environment suitability is still a concern because of the security reasons particularly for a few extremely specialised applications. But majority of HMI/SCADA systems in the market are using Ethernet TCP/IP irrespective of security risks.

However, the Ethernet TCP/IP-based SCADA systems are very much vulnerable because of high probability of cyberwarfare/cyberterrorism attacks. Since most of the SCADA systems are of critical nature, such attacks may lead to severe financial losses by loss of data or actual physical destruction, misuse or theft. It might even result in loss of life either directly or indirectly. It is yet to notice the future trend of SCADA systems of either continuing with using low-cost and highly effective Ethernet TCP/IP for communication irrespective of very high chances of vulnerable attacks or opting for more secure architectures and configurations with the involvement of significantly high cost. Multiple security vendors have already begun to address these risks by developing lines of specialised industrial firewall and virtual private network (VPN) solutions for TCP/IP-based SCADA networks. The VPN is a private communications network often used by companies or organisations for confidential communication through a public network. The VPN traffic may use a public networking infrastructure namely the Internet on top of standard protocols, or a service provider's private network with a defined Service Level Agreement (SLA) between the VPN customer and service provider. The VPNs can send data, e.g. voice, data or video, or a combination of these media, through secured and encrypted private channels between two predefined points.

7.8   Distributed control system (DCS)

The DCS is basically a control system for manufacturing system or process or any kind of dynamic system where the controller elements are distributed throughout the system with each component sub system under the control of one or more controllers unlike central controller system (CCS). Latest trend of SCADA is using DCS for communication and control. The entire DCS may be networked by communication and monitoring for use in industrial, electrical, computer and civil engineering applications to monitor and control distributed equipment with or without remote manual control leading to semi-automated or automated control paradigms.

The DCS typically uses computers with custom-designed processors as controllers with the facilities of both proprietary interconnections and open protocols for communication. The I/O modules are components of the DCS and the processor, being a part of the controller receives information from input modules and sends control command signals to output modules. The input modules receive information from input instruments of the controlled process and output modules transmit control commands to the output instruments in the field. The computer buses or electrical buses interconnect the processor and modules through multiplexers/demultiplexers as well as interface with the HMI, central control consoles and SCADA system.

The DCS is widely applied for control solutions in a variety of industries namely (i) electrical power grids and electrical generation plants, (ii) environmental control systems, (iii) traffic signals, (iv) water management systems, (v) refining and chemical plants, (vi) pharmaceutical manufacturing and (vii) sensor networks. Its broad architecture of solution involves either connections to physical equipments, e.g. switches, pumps and valves or connections via a secondary system, e.g. a SCADA system. Normally, the DCS solution does not require any manual intervention, but for interfacing with the SCADA it is necessary for skilled operators' intervention for better reliability. A typical DCS consists of distributed digital controllers capable of executing huge number of regulatory control loops in one control box, I/O devices being integral with the controller or located remotely via a field network with extensive computational capabilities in addition to proportional-integral-derivative (PID) control and logic and sequential controls. It may employ several configurable workstations, the local communication being handled by a control network with transmission over twisted-pair, coaxial or fibre-optic cables. A server with application processors may be included in the system for additional computational, data collection and reporting capability.

7.9 Sub-station communication standardisation

The IEC TC 57 (International Electrotechnical Committee (IEC) Technical Committee (TC)) was formed in early 1960s for the development of international standards in the field of communications between the equipment and systems for the electric power process and for incorporating telecontrol and teleprotection for the control of electric power systems. During the development, it was necessary to include not only equipment aspects but also control system parameters in SCADA systems, energy management systems (EMS), distribution management systems (DMS), distribution automation (DA), teleprotection and associated communications, etc. The experts identified the rising competition amongst electric utilities due to the deregulation of the energy markets and visualised the necessity of supporting the core processes of the utilities by integration of equipment and systems. It controls the electric power process to form integrated system solutions with interoperability and compatibility amongst the system components, interfaces, protocols and data models. Similar initiatives have also been taken by US-based Utility Communications Architecture (UCA) to create recommendations for implementation of interfaces, protocols and data models. The IEC TC 57 is likely to adopt these recommendations as a subset of the IEC 61850 standard. The key objective of communication standardisation is interoperability and interchangeability between vendors and systems with regard to functions, hardware and software interfaces, protocols, data models, etc.

One of the major challenges facing the development of sub-station protective relaying and metering systems is the communication capability of the component devices and keeping up with the continuous advancements and changes in communication methodologies. The main area of bottleneck and frustration of end-users is choosing the most cost-effective, low-risk and efficient option from several incompatible proprietary communication approaches and systems available in the market. Communication interfaces are mainly used to acquire data from AC voltage and current metering, power system and relay status reporting, event records and oscillographic data gathering for disturbance analysis, checking or changing the large number of settings in the flexible intelligent electronic devices (IEDs), etc. Some of the IEDs are also capable of executing basic remote control functions. In order to maintain own customer base, most of the competing IED manufacturers prefer to design their own unique approach in communication interface circuits including wide variety of types of serial ports for communicating to computers. Some of the IED manufacturers design networks to interface a number of devices in one sub-station to a single local or remote host, which is once again a unique approach that prevents the users to directly interconnect competing products. This scenario is inhibiting the end-users to avail the best service even if several efficient communication techniques are available in the market.

There is a persistent demand from the users to merge the communication capabilities of all of the IEDs in a sub-station more so in an entire power network leading to wide-area network interconnection providing data gathering and setting capability as well as remote control. Besides, multiple IEDs can participate in high-speed data sharing and control commands to execute distributed protection and control functions. This cooperative control operation can supersede and eliminate most of the dedicated control wiring as well as costly special-purpose communication channels, resulting in centralised system monitoring and control. System integrators need to employ efficient and expensive gateway or translator devices to get all the data into a common format out of a variety of communications dialects.

7.10   SCADA communication and control architecture

The SCADA communication architecture should provide to and fro communication for data acquisition and control through sub-station IEDs to facilitate LANs within the microsource sub-stations. Plentiful and cost-effective products for this huge information technology (IT) market are capable of serving LANs in the sub-station. The Ethernet, as identified by the standard IEEE 802.3, is significantly powerful and popular in the communication area. However, the requirements for IT are grossly different for business offices and sub-stations. The office IT needs to provide very few data servers and several data clients with negligible peer-to-peer communications. On the contrary, IT in microsource sub-stations needs to provide LANs only to ensure several peer-to-peer communications for supporting many data servers with only a few data clients. Besides, the microsource sub-station operating environment requires the components and devices to be quite robust to ensure reliable communications in any adverse situations namely ice storms, direct sunlight and other natural calamities. Thus in general, this sub-station IT needs to ensure security, determinism, reliability and maintainability in its communication and control architecture.

7.11   Communication devices

Some of the frequently used communication devices are briefly discussed as follows:

(1) *Hub* – The hub is a multi-port device for re-broadcasting all data that it receives on each port to all the other ports, operating at the physical layer without using any data for routing actions.
(2) *Switch* – The switch is an intelligent multiplexing device that operates at the Data Link Layer of the Open System International (OSI) network model. It is used to monitor the data received at a port in order to determine its disposition. Any incomplete or indecipherable data are ignored by the switch while the intact data packets are re-broadcasted to other suitable ports. Some switches can operate on the Network Layer or Transport Layer packet information.

BMU, BPUs and other IEDs instead of, or in addition to, the IEDs communicating directly through the bay and station Ethernet switches to the Station Control Unit (SCU). Although IEC 61850 was previously designed specifically for IED quantities associated with sub-station control, current requirements call for the conversion of the IEC 61850 to legacy telecontrol protocols suitable for communicating to the remote control centre (RCC). Presently, sub-station LAN traffic is so high that LAN segmentation is necessary. Therefore, even if high-speed Ethernet connections between the RCC and sub-station are established, the RCC may not be capable of handling the process burden of communications directly to each IED. In that case, SCUs or BCUs must act as single points of contact to communicate data and control.

(4) *Interoperability and interchangeability amongst clients and servers* – The IEC 61850 standard is of significant use in designing bay LANs in combination with IEDs. The required IEC 61850–IED interface can be developed by vendors up to customers' satisfaction if the end-users can provide the IEC 61850 communication requirements. This implementation may not be sufficient to satisfy other end-users with different requirements asking for modification for each situation. Unmapped existing data within server IEDs need to be exposed to the IEC 61850 LAN during extension or augmentation of LNs. Integration efforts necessary for new operator interfaces, logic and control are significantly reduced with availability of standardised names and attributes. However, during the replacement of one bay LAN by another one containing different LN and GSE characteristics, reconfiguration of the database, logic and operator interfaces is necessary. Some end-users opt for interchangeability, i.e. replacement of any IED within one bay by another IED from any manufacturer. It may result in affecting the function of the co-ordinated system because of the use of different operating principles of different vendors. Hybrid LANs are capable of supporting interchangeability with many different IEDs at the IEC 61850 communication interface level creating appropriate LN and GSE interfaces in the new IED or within server IEDs on the LAN.

(5) *System implementation* – Communications traffic on actual sub-station Ethernet LANs will not be limited to the protocols included in the IEC 61850 standard. In future, sub-station LANs will also be able to support traffic for web server applications in the IEDs and HMIs, non-standard vendor-specific IED protocols, Modbus IP and DNP IP, e-mail, legacy SCADA protocols such as IEC 60870, vendor-specific IED configuration and diagnostic applications, network analyser configuration and other diagnostic applications and telephone and camera applications. During implementation, it must ensure that the IEDs support the necessary data requirements and that their performance characteristics are in compliance with the other components to create a successful network. In future, it is likely that a flexible data mapping technology at the IED and bay level will make the physical application of an IEC 61850 network a much easier task.

7.13   Conclusion

The overall control and management of a Microgrid will have to be implemented through an intricate network of state-of-the-art IED devices interlinked through SCADA and high-speed communication channels. The central controller for Microgrids will perform functions like energy management, management of ancillary services, metering and protection co-ordination, grounding co-ordination, inter-tripping and fast tripping of circuit breakers through its Energy Management Module (EMM) and Protection Co-ordination Module (PCM). The entire operation of these modules will depend on high-speed communication and interoperability between the devices. Therefore, the development of communication standards for microsource sub-station automation with SCADA will have immense significance in designing, developing and implementing EMM and PCM for a Microgrid and an active distribution network. Standardised communication protocols will also help in the integration of several Microgrids to form a quality power park. The power quality and reliability of Microgrids and active distribution networks are discussed in Chapter 8.

Chapter 8
Impact of DG integration on power quality and reliability

8.1 Introduction

Modern electrical distribution systems are complicated aggregation of several components and numerous supply points. Their interaction with the power utility results in temporal variations in the characteristics of the power supplied to the customers. These variations usually appear in the form of very short to longer periods of outages or abnormal voltage and/or frequency characteristics. The quality of the supplied power is dependent on these variations, whereas reliability depends on the frequency of interruptions and outages.

Because of the introduction and widespread use of several sensitive electrical and electronic gadgets in commercial and industrial sectors, power quality and reliability issues have gained considerable importance in recent years. In order to safeguard the sensitive systems from the detrimental effects of power quality and reliability problems, several customers are investing towards procuring and installing custom power systems for conditioning and supplementing supplied power. This has led to the development of the premium power market for dealing with manufacturing, sales, procurement and installation of such power-conditioning equipments.

Distributed generation (DG) and integration of distributed energy resources (DERs) in the form of Microgrids can be used to improve power quality and reliability significantly to suit the needs of the customers. The potential services that can be provided by DGs and Microgrids are as follows:

(1) The application of combined heat and power (CHP) systems help to enhance the overall energy efficiency of the power system. Moreover, the combined use of thermal and electrical energy can make CHP a more economic option for customers than buying electricity and fuel for thermal loads separately.
(2) The generation of power locally with renewable or non-conventional energy sources like landfill gas, biomass/biofuels or photovoltaic (PV) systems (with or without heat recovery systems) becomes much more cost-effective to customers who are remotely located from the central generating stations.

Over-voltages, on the contrary, may occur due to problems with voltage regulation capacitors or transmission and distribution transformers. The problems are magnified when the over-voltage protection devices do not respond fast enough to completely protect all equipment downstream. Over-voltage problems are usually eliminated by installing voltage regulator devices at key distribution sites within the customers' premises such as the service entrance, the main distribution panel or the computer room panel or by installing UPS systems both to regulate the voltage to sensitive loads when utility power supply is available and to provide backup power in case of utility supply failure. Power utilities are designed to maintain voltage ranging from +10% to −10%, and are also provided with adequate over- and under-voltage and frequency protection systems to safeguard their equipment from abnormal voltage and frequency deviations beyond the specified range.

8.2.4   Outage

Outage or voltage interruption refers to the complete loss of voltage over a certain period of time. Outages may be short term (less than 2 minutes) or long term. These are normally caused by the opening of an isolating device (circuit breaker or line recloser) or by a physical break in the line. In case of any fault in a transmission or distribution feeder, the circuit breaker or recloser will immediately open in an attempt to clear the fault and the customers connected to the faulted feeder will experience one or more interruptions, depending on the type of fault and reclosing practices of the power utility. Temporary faults are usually cleared after one or two reclosing attempts and the normal supply is restored whereas for permanent faults, the circuit breaker locks out after a set number of reclosing attempts, resulting in a longer-term outage on that line. Customers on that line will experience sustained outages and those on parallel lines will experience voltage sags during the fault and subsequent reclosing attempts.

Outages to a system can be alleviated by installing UPS systems with battery storage and power-conditioning equipment, by storing mechanical energy in large high-speed flywheels or by arranging for multiple feeds to the facility. Protection from momentary interruptions however requires a static source transfer switch (SSTS). Protection from sustained outages (beyond the energy storage capacity of UPS or battery systems) can be provided by on-site generation from diesel generator sets or low emission distributed generators based on non-conventional/renewable technologies.

8.2.5   Harmonic distortion

Harmonic distortion arises when the shape of voltage or current waveform deviates from the standard sinusoid. Harmonic distortion implies that apart from standard power frequency component, higher-frequency components are also present in the power flow. These components can degrade equipment performance and may even cause damage to it. Some possible problems caused by harmonics are overheating of distribution transformers, disrupting normal operation of electronic equipment and system resonance with power factor correction banks. Potential sources of harmonics may be computers, lighting ballasts, copiers and variable frequency drives. Harmonic disturbances may be avoided or controlled by using equipment like 12-pulse input transformer configuration, impedance reactors or passive and active filters.

8.2.6   Voltage notching

When silicon-controlled rectifiers (SCRs) are used in electrical control systems, line voltage distortion in the form of 'notches' may occur in the waveform. Line notches typically occur in the waveform either during SCR commutation or when a single-phase SCR is turned off and the next one is turned on. During this small period of time, a momentary short circuit exists between the two phases, resulting in the current rising and the voltage dropping. This appears as a notch in the voltage waveform. The most severe and damaging form of notch is the one that touches the voltage zero axis. The types of equipment that frequently use SCR control schemes and experience notching include DC motor speed controls and induction heating equipment. A voltage waveform with a typical line voltage notch is shown in Figure 8.1.

Proper functioning of various electronic equipment is based on the detection of zero crossing in the voltage waveform. Some equipment need to be triggered at the zero crossing in order to avoid the possibility of any surge currents or inrush currents while some, like digital clocks, use the zero crossing for an internal timing signal. Notches touching the zero voltage axis may appear to be a zero crossing to such equipment, thereby causing them to malfunction. Sensitive equipment connected to the same voltage source as the equipment producing the notching can be protected by installing a 3% impedance reactor which eliminates multiple zero crossing and mitigates interference with neighbouring equipment.

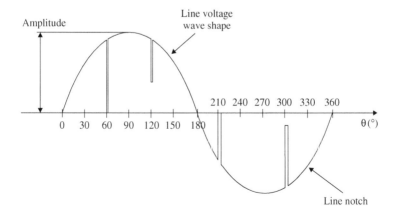

Figure 8.1   Voltage notching

8.2.7 Flicker

Flicker is defined as a modulation of the voltage waveform at frequencies below 25 Hz, detected by the human eye as variation in light output from standard bulbs. Voltage flicker is normally caused by arcing on the power system from welding machines or electric arc furnaces. Flicker problems can be eliminated by installing filters, static VAR systems or distribution static compensators.

8.2.8 Electrical noise

Electrical noise is defined as a form of electromagnetic interference (EMI) caused by high-frequency, low-voltage signals superimposed on the standard signal in a line. Frequencies of these signals may vary from the range of kilohertz to megahertz while magnitudes may be up to 20 V. EMI adversely affects telecommunication processes and hence is called noise. It arises from a variety of natural and artificial sources like lightning, static electricity and solar radiation, presence of power frequency transmission lines in the vicinity, automobile ignition, high-frequency switching in power electronics devices and fluorescent lamps. Equipment that are adversely affected by noises are computers, industrial process controls, electronic test equipment, biomedical instruments, communications media and climate control systems. The impact of noise may be reduced by installing radio frequency line filters, capacitors or inductors at the equipment level.

The impact of power quality disturbances on equipment operations depends not only on the type of electricity-using equipment in place but also on the frequency of occurrences throughout the year. Even when the duration of typical voltage sag or interruption is very brief, the impact on customers varies widely depending on the voltage or frequency sensitivity of the equipment. Most sensitive customers may be adversely affected for several hours. Studies have shown that almost half of disturbances are voltage sags/swells while the next most common problem is harmonic distortion followed by wiring/grounding problems at the facility. It should, however, be remembered that many power quality problems do not result from the utility system practices rather these may arise from the customer's own power-using equipment or the power use of a neighbouring customer.

8.3 Power quality sensitive customers

With the advent of sophisticated voltage- and frequency-sensitive electrical and electronic gadgets, supervisory control and data acquisition (SCADA) systems and computerised process control systems in commercial and industrial sectors, power quality and reliability issues have currently gained considerable importance as a major criterion for judging quality of service. In order to safeguard the sensitive systems from the detrimental effects of power quality and reliability problems, several customers are investing towards procuring and installing custom power systems for conditioning and supplementing supplied power. This has led to the development of the premium power market for dealing with manufacturing, sales, procurement and installation of such power-conditioning equipment like UPS systems, battery and flywheel storage systems and diesel generator sets. The possibility of using environment-friendly DG technologies like microturbines, solar and wind power, fuel cells for improving power quality and reliability, either as stand-alone systems or integrated in the form of Microgrids, are also being explored.

As discussed in the Section 8.2, power quality disturbances occur when the voltage waveform supplied by the utility fails to conform to the standard sinusoidal wave shape of constant amplitude and frequency. The ultimate form of power quality disturbance is sustained outage or complete loss of voltage at the customers' terminals. The economic impact of these outages varies widely according to class of customers or according to the sensitivity of equipment used by a specific customer. Customers, who cannot afford to be without power for more than a brief period, usually install on-site standby generator sets to provide backup power to the priority loads during supply interruptions. On the contrary, the customers who have to suffer from severe economic losses due to any disruption of supply or variation in power quality generally install UPS systems along with associated power and conditioning equipment to eliminate the effects of surges, sags, harmonics and noise.

Customers with a need for true premium power systems include the following:

(1) *Mission critical computer systems* – Banks, depository institutions, financial companies, stock markets, investment offices, insurance companies, computer processing companies, airline/railway reservation systems and corporate headquarters that need to protect their computers, peripherals and computer cooling systems.

(2) *Communications facilities* – TV/radio stations, telephone companies, Internet service providers, cellular phone stations, repeater stations, military facilities and satellite communication systems that must protect their computers, peripherals, antennae, broadcasting equipment and switches.

(3) *Health care facilities* – Health care facilities like hospitals and nursing homes that need support to maintain critical life support systems, medical equipment and ensure proper maintenance of critical heating, ventilation and air conditioning (HVAC) environments.

(4) *Large photofinishing laboratories* – These must protect their computers and photofinishing equipment.

(5) *Continuous-process manufacturing systems* – These manufacturing systems in paper, chemical, petroleum, rubber and plastic, stone, clay, glass and primary metal industries for whom any supply interruption would result in loss of production.

(6) *Fabrication and essential services* – These services and all other manufacturing industries plus utilities and transportation facilities such as railroads and mass transit, water and wastewater treatment, and gas utilities and pipelines.

to damages by high currents and overheating of customers' equipment. Voltage fluctuations also tend to reduce the life expectancy of equipment. Therefore, following reactive power compensation technologies are employed to restore and maintain voltage stability:

  (i) Synchronous condenser
  (ii) Fixed capacitors banks placed near large inductive loads
  (iii) Thyristor-switched capacitor (TSC)
  (iv) Thyristor-switched reactors (TCR)
  (v) Static VAR compensator (SVC)
  (vi) Static synchronous compensator (STATCOM)
  (vii) Active VAR compensator

(3) *Dynamic voltage restorer* – DVR provides the system with adequate buffering to ride through temporary disturbances such as voltage transients like dips, sags and swells. It is connected in series between the grid and protected load and stabilises the voltage at the customer's systems during transients caused by faults in the transmission or distribution system. The DVR can be designed for any voltage and load requirements but is best suited for medium- and high-voltage applications as in case of industrial and large commercial customers. Energy storage for up to 300–500 ms sag is provided by the capacitor banks in DVR.

(4) *Isolation transformer* – Statistically shielded isolation transformers are used to shield sensitive loads from EMI. Such loads include sensitive electronic and computerised equipment used in medical and surgical rooms or in very precise process controls. The isolation transformers protect these loads against indirect contacts without interrupting the circuit upon an initial ground fault. Hence, these are preferred for installations where a sudden interruption of load is undesirable and automatic interruption is strictly prohibited.

(5) *Motor–alternator sets* – Motor–alternator technology is used as a very efficient 'line conditioner' providing both voltage stabilisation and noise rejection. The set consists of an AC or DC motor coupled to a generator or alternator that supplies power to the priority loads during power failures. It not only rejects common-mode noise, but also prevents any line-to-line noise from entering the output because of the shaft or belt connection. The rotary UPS systems use motor–generator sets with their rotating inertia to ride through brief power supply interruptions. In this system, the generator provides true isolation of the power so that no abnormalities pass through the UPS apart from some slight harmonics produced by characteristics of the generator's windings.

(6) *Uninterruptible power supply* – In the double-conversion or premium UPS system, the AC input power is rectified to DC power to supply the internal DC bus of the UPS. The output inverter converts the DC power to regulated AC power at standard power frequency to supply the priority load. During normal operation (when grid power is available), batteries attached to the DC bus are float charged while during grid power failures, the

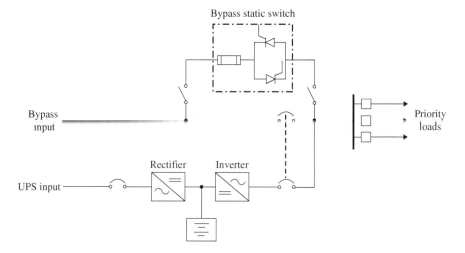

Figure 8.2   Scheme for double-conversion UPS system

batteries provide power to the DC bus to support the inverter and the priority loads. The configuration of a double-conversion UPS is shown in Figure 8.2.

This UPS system consists of the following subsystems:

(1) *System controls* – The system control logic automatically manages critical bus operation and monitors performance of the UPS module providing interactive display and port communication with external devices through microprocessors and dedicated firmware.
(2) *Rectifier/charger* – The rectifier/charger converts utility power from AC into DC to charge the battery and provide the DC input to the inverter with low ripple DC power preventing harmonic current distortion in source power.
(3) *Inverter* – The inverter converts DC power into the precise AC power required to supply a sensitive priority load. It converts DC power into a pulse-width-modulated (PWM) waveform with easy filtering to produce a clean sine wave output minimising the harmonic voltage distortion caused by typical switching power supplies and other non-linear load components.
(4) *Static bypass switch* – It is solid-state bypass switch that transfers the load quickly from the inverter to the bypass AC power source during a severe overload on the system or a failure within the UPS. It takes place without any interruption of power supply to the load. The system needs to include redundant circuits to detect and isolate shorted SCRs in the static bypass switch.
(5) *Fuses* – These are installed in series with the static bypass circuit for reliable overload protection in case of any catastrophic output condition. The static switch SCRs are sufficiently rated to handle the fuse-blowing current.

(6) *Bypass circuit* – It consists of a motor-operated circuit breaker in parallel with a solid-state switch and associated synchronising and control circuitry to transfer the load to/from the bypass source.

(7) *Battery energy storage system* – It is used as the alternate source of DC power supply to the inverter when AC supply voltage is outside the acceptable range. The battery supplies power to the inverter until the utility power is restored or an alternate power source is available. If AC source power is not restored or an alternate power source is not available, the battery can be sized to provide power long enough for an orderly shutdown of the load.

The major advantages of this UPS system are as follows:

(1) The priority load is completely isolated from the incoming AC power.
(2) The priority load is always being supplied by the inverter, which is always being fed via the internal DC bus. Thus, in case of input power failure, there is no transitional sag in the output voltage as the inverter is already operating on DC input.
(3) Even if the AC input voltage and frequency fluctuate, the double-conversion UPS does not notice it, since the rectifier is only making DC power to feed the DC bus. The UPS can operate and even continue to recharge its batteries with input voltage at 15% below nominal. It can continue to operate, without discharging the batteries, through voltage sags of 20% below nominal. Likewise if input frequency is fluctuating in and out of specification, the rectifier will continue to produce DC power and the output inverter will continue to produce 50 Hz power without using the battery.
(4) The output inverter usually contains an isolation transformer producing a separately derived neutral. This enables the UPS to be electrically isolated, providing common-mode noise protection for the load.
(5) The double-conversion UPS is inherently dual-input, i.e. having separate inputs for the rectifier and bypass circuits.
(6) A fault on the input line causes the UPS to go to battery power, but the UPS rectifier will not allow power from the DC bus to flow upstream.
(7) This is a very well understood design with a long track record of proven performance. Though battery energy storage is the most common form of storage used in a UPS, other forms of energy storage are being used and/or developed for commercial use. These emerging systems include flywheels, super capacitors and superconducting electromagnetic storage, etc.

8.5 Impact of DG integration

Premium power market reliability and power quality can definitely be improved by the use of DG and its integration. Reliability and power quality improvements are major impacts of DG integration in the form of Microgrid through active distribution network. The following requirements are the main elements of power quality and reliability improvement:

(1) *Fast response* – The load needs fast response of energy storage system to safeguard from momentary voltage fluctuations.
(2) *Clean power* – Storage power should be converted to clean power.
(3) *Synchronisation* – Smooth control should be there in paralleling and synchronisation.
(4) *Soft transfer* – Alternate power source should be capable of seamless power transfer.
(5) *Isolation* – Integrated DG, i.e. Microgrid should be efficient enough in quick isolation from utility in case of any contingencies.
(6) *Adequate storage* – The quantity of storage energy should be sufficient to ride through any outage until primary or secondary power restoration.
(7) *Supply to priority loads* – Microgrid should be capable of supplying clean power to the priority loads of the system.
(8) *Dispatch ability* – Microgrid should be capable of supplying power to varying local loads.
(9) *Efficiency* – Microgrid should operate at high efficiency.
(10) *Emission* – Microgrid should drastically reduce DG emissions to minimise the environment impacts.

Some examples of DG integration schemes are as follows:

(1) Simple standby generation scheme
(2) Secondary DG system with power quality support
(3) Primary DG system with power quality support to priority loads
(4) Soft grid-connected DG with power quality support to priority loads
(5) DG with intermittent solar PV within power quality environment
(6) DG with intermittent wind generator within power quality environment
(7) Ultra-high reliability scheme using dual link DC bus

8.5.1 Simple standby generation scheme

A standby generator safeguards the loads from long-term outages as shown in Figure 8.3. It is connected to loads through automatic transfer switch (ATS). The logic controller automatically senses any utility grid outage and changes over the load through ATS and starts the generator.

A typical diesel generator takes 10 seconds for supplying power to loads. The loads are prioritised to maintain supply continuity to the priority loads without generator overloading. It can safeguard loads only from any long-term outage without protecting loads from any short-term voltage disturbances. Therefore, this scheme is not that effective for the customers sensitive to short-term power quality. The use of uncontrolled diesel power generator is restricted for its pollution emissions and switching control limitations. However, environment-friendly alternative fuel generators with sophisticated switchgear may be widely used in this scheme.

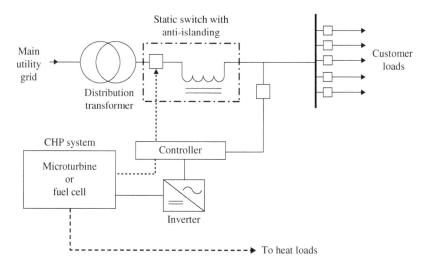

Figure 8.6   Soft grid-connected DG with power quality support to priority loads

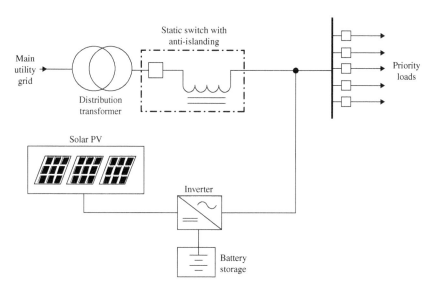

Figure 8.7   Solar PV as intermittent DG within power quality environment

8.5.6   DG with intermittent wind generator within power quality environment

Figure 8.8 shows the configuration of connecting renewable DG with intermittent supply from wind turbine generator.

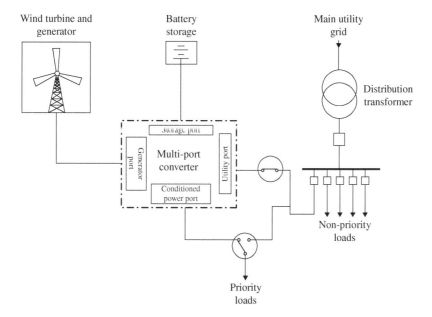

Figure 8.8 Wind turbine as intermittent DG within power quality environment

This configuration supplements energy consumption with adequate support for on-site power quality requirements.

8.5.7 *Ultra-high reliability scheme using dual link DC bus*

Figure 8.9 shows a typical ultra-high reliability scheme using dual link DC bus. This system is used by customers who require extremely high reliability and it operates independent of utility grid. It integrates on-site power generation, UPS, flywheels connecting through dual bus with more than sufficient redundancies for maintaining continuity of high-quality power supply.

In this scheme, DG is used as the main power supply and utility grid as a backup supply without any grid connection during normal operation. Flywheels protect against any DG outage as well as any step load change. Waste heat from continuously running DG system is utilised for HVAC applications.

This scheme has the following advantages:

(1) Source voltage failures do not draw power from other sources.
(2) Priority load faults are isolated to UPS motor generator since the fault clearing is superior to UPS.
(3) No system failure takes place due to control system failure.
(4) Multiple generators operate independent of each other.
(5) There is no reverse power flow.
(6) Synchronisation or cascade failures are remote.

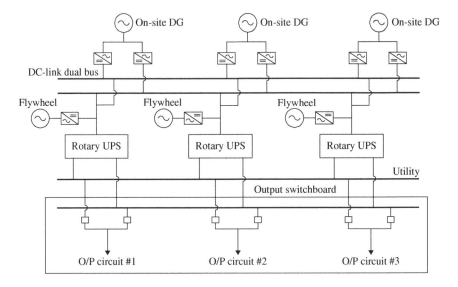

Figure 8.9   Ultra-high reliability scheme using dual link DC bus

8.6   Issues of premium power in DG integration

The integration of DG with UPS and other premium power devices needs to address the following issues for proper functioning:

(1) On-site storage capacity should be sufficient to sustain till the generator starts up to supply. A diesel generator may be started within 10 seconds whereas CHP microturbine may take up to 90 seconds. Therefore, flywheels with sustaining period of less than 30 seconds may be used as backup of diesel generators not of microturbines.

(2) The DG system support for priority loads with motor starting or other step loads should be sufficient to avoid any loss of voltage control. DG systems with fuel cells, microturbines may not be capable of handling step load changes unlike diesel generators.

(3) DG systems operating in parallel with utility need to have necessary interconnection protections for positive disconnect/reconnect, voltage/current regulation, etc. in fault clearing and reclosing technology.

(4) DG systems with UPS and/or standby generators should take care of synchronisation problems for constant temperature HVAC controls wherefrom many detrimental load pulses may be generated.

(5) DG systems with UPS using static filters should take care of excess capacitance on light load condition, which may result in loss of voltage control by power factor mismatch. It should be capable of handling quality power loads.

(6) DG systems should provide sufficient redundancies for supplying extremely high reliability power to sensitive loads using dual bus, ATSs etc.

(7) DG systems need to be separated from each other by fire rated walls to protect system loads during any fire hazards.

(8) DG systems in standby mode need to have dual starter, special fuel filters etc.

(9) DG systems used as prime power supplier need to fulfil all the aspects of power quality and reliability.

8.7 Conclusion

Customers are more eager to invest for power quality and reliability to safeguard sensitive loads from subtle and frequently occurring supply disturbances than from any blackout that might take place rarely. Several power premium devices like power factor correction controllers, harmonic filters and isolation transformers are used in addition to centralised as well as isolated UPS systems. Standby diesel generators, CHP-based microturbines are used in different schemes as explained earlier. Microturbines with heat loads may be used either as primary generation in parallel with utility grid for economy in generation or as stand-alone backup generation with anti-islanding protection for emergency protection. Normally, utility services ignore repeated customer requests for betterment of power reliability and power quality leading to customers to install power monitoring devices to capture on-line data for identifying and diagnosing power quality problems for getting the power quality problems addressed by utility services. It has been identified that DG integration can be fruitfully used to support customers' power quality and reliability requirements. Peak shaving CHP applications in DG integration may play a major role in reducing capital costs by alleviating any investment in standby diesel generators.

CHP applications in DG integration may be very useful to address power quality problems of less sensitive loads because of customer's reluctance to invest in UPS. Critically power quality sensitive customers are not interested to get this benefit of DG integration because of their existing dedicated UPS systems.

In order to make DG integration a potentially widespread solution for power quality and reliability problems, the following issues need to be addressed:

(1) Both standby generators and DG systems themselves may cause power quality problems due to system incompatibility of different electronic devices. Hence for any future installation and commissioning of DG system, it needs to be correctly designed to address incompatibility issues.

(2) Interconnection rules, which prevent DG systems to provide backup power during utility grid outages, need to be re-examined from both technical and economic perspective.

(3) Customers having own DG systems need to be encouraged in supplying emergency demand of the integrated system.

Microgrids. The accumulated knowledge of power system operation in grid scale needs to be optimally applied in the distribution level grid, i.e. the Microgrid. The established and reliable tools should be suitably applied in Microgrid operation.

(2) *Utilisation of unique aspects* – Some of the unique aspects of Microgrid economics need to be utilised properly. Unlike conventional distribution systems, Microgrids can provide heterogeneous levels of reliability to the end-users as per necessity of the customers. The operational constraints of centralised power system economics might not be similar to that of Microgrid economics. For example, the constraints on generation of noise level are relatively insignificant to centralised power providers unlike Microgrids.

(3) *Distribution system relationship* – The relationship of Microgrid with the distribution system is an important aspect of Microgrid economics. Microgrid needs to have real-time price signal in successful interface between customers and utilities. It helps to achieve optimal use of resources by both Microgrid and utility grid. It should be noted that the ability of Microgrid to participate in utility grid-scale ancillary service markets is limited in supporting voltage and losses. Nevertheless, Microgrids can provide excellent local ancillary services in voltage support and others for the end-users. Microgrid economics would be drastically improved by providing its market participation for localised voltage support and other ancillary services.

9.3   Microgrids and traditional power system economics

Microgrids are normally designed and commissioned for operation by a group of customers keeping an eye to minimise environmental impacts. However, the main target is to minimise the total energy bill for heat and electricity of the participant customers. Microgrid can supply energy at lower cost because of its optimal use of waste heat. Besides, Microgrid does not have any cost for transmission and distribution losses, customer services, congestion and other allied costs unlike traditional power systems. Microgrid has several advantages to reduce its energy costs compared to traditional power system. On-site generation is competitive with conventional power generation particularly with reciprocating engines. However, its environmental impact and interconnection costs sometimes restrict its applicability. The emerging DER technologies are quite promising in generating low-cost clean energy. It must be mentioned that latest carbon capture and storage (CCS) technology is on board for application in existing power plants to generate clean electricity with higher energy efficiency in the tune of about 50%.

DER technology can challenge the economics of central power generation in addition to potential ancillary benefits of DER. But the conventional system offers low risk as well as low transaction cost to the end-users. Major hindrance of Microgrid economics is its hidden overrun and subtle costs in installation and commissioning. High installation costs of Microgrid DERs are subject to government subsidies particularly to install efficient storage systems for its islanding mode of operation. However, the installation cost should be reflected in the economic evaluation of Microgrid. The additional costs need to be considered against the added customer benefits from islanding capability as well as the additional utility grid costs of maintaining high system reliability. Simple engineering–economic principles can be applied to determine suitable DER technologies and their deployment modalities for Microgrids.

Economics of Microgrids have several similarities to those of utility grids in the following aspects namely (i) rules of economic dispatch, (ii) cost minimisation by lowest-possible-cost combination of resources as per equipment characteristics, (iii) purchase and sale of electricity occurring at different times, (iv) optimal combination of technologically diverse resources to suit various duty cycles of the system, (v) suitability of high capital but low variable cost generator technologies for catering base loads and (vi) suitability of low capital cost but high variable cost generator technologies for catering peak loads.

Economics of Microgrids have some major dissimilarities to those of utility grids in the following aspects:

(1) Joint optimisation of heat and electric power supply
(2) Joint optimisation of demand and supply.

9.3.1 *Joint optimisation of heat and electric power supply*

CHP systems are an underdeveloped area of Microgrid economics. In some countries, CHP systems are partially used for generating electricity. Nevertheless, its main objective is to utilise the waste heat energy for heating appliances, in addition to the production of electricity by joint optimisation of heat and electricity. In the economics of central generating systems, the use of heat was never the central objective. The reason for rethinking on the use of CHP systems is drastic reduction in carbon emissions and increase in overall power generation efficiency. The overall generation efficiency can be increased from conventional 33% or combined cycle gas turbine (CCGT) 50% to more than 80% by using CHP systems. CHP is the heart of Microgrid economics. Maximum energy efficiency can be achieved by minimising heat transmission losses between generators and loads. Hence, CHP-based generators are located near the premises of the heat loads. The major applications of CHP in Microgrids are namely (i) space heating, domestic water heating and sterilisation; (ii) heating for industrial/manufacturing processes; and (iii) space cooling, refrigeration through absorption chilling. The technical feasibility of exploiting CHP opportunities is quite self-motivating for customers to generate CHP-based electricity. It is quite cheaper for end-users compared to the purchase of electricity, heating and cooling services separately from different sources. Besides, CHP is the main motivation for multiple customers to join together to install and commission CHP-based Microgrids. However, comprehensive economic studies need to be performed on the CHP opportunities to examine aggregating benefits of heat and electric loads in order to facilitate multiple customers to form Microgrids.

9.3.2   Joint optimisation of demand and supply

Joint optimisation of demand and supply is the second priority for Microgrid economics. Traditional power economics need to be extended for Microgrids. In utility grid-scale economics, load control is addressed in the analysis and planning stages in the form of demand side management (DSM), load shedding, interruptible tariffs/contracts, etc. In case of Microgrid, the most important criterion of this optimisation is the marginal cost of self-generation at any point of time. Investment cost recovery, cross subsidies, inaccurate metering and tariffs are not considered in the generation economy. In utility grid-scale economics also tariffs and environmental aspects are not well represented. The demand and supply optimisation in Microgrid is easier since here the generator and consumer is one and the same decision-maker. The Microgrid should know both its marginal cost of power generation at any point of time and the equivalent costs of investments in energy efficiency. It can easily decide the cost of curtailment for trading. It leads to implementation of a new paradigm in load control.

9.4   Emerging economic issues in Microgrids

There are some Microgrid economic issues covering unique Microgrid features unlike traditional utility grid-scale economics. Microgrids can provide heterogeneous levels of reliability to various end-users within the Microgrid. Microgrids unlike central power generation need to operate at constraints like reducing noises by generators. The design and operation of traditional power systems are normally to deliver power to all customers at uniform power quality and reliability irrespective of different needs of the end-users. But Microgrids control the supply power quality and reliability at the points closer to the end-users resulting in serving the customers to their satisfaction maintaining power supply at heterogeneous levels of reliability as per the customers' needs. Thus, low-end customers with power demand of lower quality and reliability can save their energy bills by procuring energy at cheaper rates. Similarly, high-end customers with power demand of higher quality and reliability can safeguard their sophisticated appliances by paying higher rates for energy supply without any extra investment on premium power supply equipments. Besides, the Microgrids can supply high-end customers in case of any shortfall of generated energy by shedding loads of low-end customers. Widespread application of Microgrids, with locally controlled generation, backup and storage with DSM, can effectively serve sensitive loads, thus resulting in the reduction of economic burdens of utility grids in maintaining uniform optimal level of power quality by constrained reliability requirements of sensitive loads.

9.5 Economic issues between Microgrids and bulk power systems

There are some economic issues of Microgrid in its relationship with the utility grid. Basic Microgrid paradigm is its co-existence with the existing utility grid as a citizen. Microgrid needs to adhere to the utility grid rules applicable to all connected devices. In the utility grid perspectives, Microgrid behaves as a cluster of customers or generators or both resulting in extended traditional economic rules. In order to accommodate growing loads, it is necessary to augment distribution systems that are not straightforward in case of Microgrids, because generators are normally embedded within the existing radial distribution system. Hence, the price signal delivery to the new customers is a bit complicated. Price signals can be delivered to the new customers during congestion in a form suitable to encourage Microgrid development and investment in augmented generation as well as load control for mitigating congestion. However, its implementation is difficult because of the dependencies on neighbouring network configurations. In a densely populated area, any end-user might have several options to be fed from existing neighbouring distribution networks to choose as per available economic signals. Thus, the congestion costs seen by any Microgrid is dependent on the available neighbouring distribution systems configurations that can change abruptly, disrupting the Microgrid economics dependency in the locality.

Microgrids need to have full participation in both energy and ancillary services market, but its low voltage level inhibits its ability to deliver power and services beyond sub-station. However, the advantage of on-site generation of Microgrids can readily be utilised with suitable control and protection schemes to provide reliable supply to the sensitive loads. This is a valuable contribution in the overall power system health and economics, because the market responses to rapidly variable load changes are practically infeasible.

9.6 Microgrid economics: the UK scenario

In Microgrid economics, the reduction of greenhouse gas (GHG) emissions is one of the most important contributions. Micro-CHP-based generation is mainly focused after the formation of distributed generation co-ordination group (DGCG) in the UK in early 2000. The potential for widespread adoption of Microgrid technologies is identified. The barriers that have to be overcome to make Microgrid a significant contributor to UK energy system are also identified. Some major economic issues need to be addressed for large-scale deployment of Microgrids in the UK. These are discussed briefly in the following sections.

9.6.1 Microgeneration

The prospects of small-scale gas-fired Stirling engines are enormous to provide domestic heat supply in addition to electricity supply by domestic-scale micro-CHP installations. The household heat requirements indicate that the by-product electricity would be surplus to be fed into national electricity grid.

(iii) Provision of power quality improvement
(iv) Provision of voltage support
(2) Ancillary services
   (i) Provision for reactive power support not necessarily for voltage
 (ii) Provision of frequency response
(iii) Provision of reserve
(iv) Provision of black start.

9.6.1.4   Market potential issues for micro-CHPs

The following critical issues are identified that need to be properly addressed for making the application of micro-CHP on a large scale in the UK:

(1) Grid connection modalities for participation in a mass market
(2) Product testing and field trials
(3) Proper delivery chain with skilled workforce for installation and service
(4) Value maximisation of micro-CHP generation by simplified metering and proper trading procedures
(5) Review of 28-day utility rule to enable energy providers to undergo long-term contracts with the retailers.

9.6.2   Regulatory issues and regulation activities

It is observed by DGCG that the UK electricity market is not friendly to the large-scale adoption of microgeneration. In addition to high installation costs, Microgrids have to face several barriers in open market participation. Some of the key issues that need to be addressed are as follows:

(1) *Deep connection charges* – Microgenerators need to pay deep connection charges unlike large-scale generators.
(2) *High market participation cost* – Microgenerators are to pay high charges for market participation in order to supply power to the existing utility grids. In order to encourage household micro-CHP installations, standardised generation connection with proper supply agreement should be undergone.
(3) *Metering facility* – The distribution network has unidirectional metering facility without any provision for bidirectional metering for demand and supply.
(4) *DNO incentives* – DNOs do not get any incentive for connecting small generators.

Ofgem is in the process of considering changes to the regulatory regime for addressing charging issues and price control mechanisms to encourage microgeneration.

9.6.3   Microgeneration technologies: economic perspectives in the UK

It has been observed from the UK economics of microgeneration that PV arrays have a payback period of about 35 years compared to about 14 years for

Stirling engine-based micro-CHP and about 6 years for household gas boilers. Thus, the replacement of central heating boiler by Stirling engine-based micro-CHP is significantly beneficial from the economic perspective.

Microgeneration suffers from tax rule discrepancies without any capital allowance unlike corporate sector. These discrepancies need to be tackled properly for wide application of micro-CHPs. The subsidy scheme of Department of Trade and Industry (DTI) should be suitably used in microgeneration. Renewable Obligation Certificate (ROC) revenue can contribute to Microgrid economics. Microgeneration can redefine supplier–consumer relationship. Distributed control systems can be utilised for automatic switching control of home appliances as per price signals and consumer preferences. This may inculcate the sense of shared responsibility leading to a sustainable energy system.

9.6.4 *Potential benefits of Microgrid economics*

Microgrids have the following significant economic benefits:

(1) Potential reduction in transmission and distribution costs and energy losses.
(2) Significantly high total energy efficiency.
(3) Significant reduction in capital exposure and risk by small-scale individual investments and closely matching capacity increases to demand growth.
(4) Lower capital cost helps to have low-cost entry in open competitive market.
(5) Microgrid microgenerators with very high energy efficiencies can share energy by the consumer generators within the Microgrid itself without any necessity to export energy to the public network at lower prices. It enhances Microgrid economics.
(6) Additional security and ancillary services from DG are special advantages of Microgrid.

The essence of economic scenario is financing for longer terms rather than short terms. In most of the renewable energy resources, the initial capital cost is predominant with a minimum running cost. Thus, the cost of consumer is dependent on financing terms and repayment schedules. In case of PVs, longer-term financing is in progress leading to an economic benefit to the consumers.

The UK-centric Microgrid economic scenarios might be categorised as follows depending upon degrees of integration of DGs and combination of different DER technologies:

(1) *Scenario A* – Distributed CHP units are installed separately without Microgrid integration. No PVs and no storage devices are incorporated in this to gauge the value addition by Microgrid integration. Economic assessment is done on the basis of micro-CHP household energy production with a low ratio of electricity to heat output without any integration to

Microgrid. In the export tariff arrangement with electricity supply company, net metering may or may not be installed.

(2) *Scenario B* – Microgrids integrate distributed CHPs and battery storage devices for supplying both electricity and heat in winter. The running of micro-CHPs is dictated by household heat requirement and hence during summer the electricity shortfall is consumed from the local utility supply.

(3) *Scenario C* – The Microgrid consisting of micro-CHP, PV and storage is capable of stand-alone operation. The PV installation subsidy varies from 50% to the full cost.

The reduction in household electricity cost by installation of micro-CHPs becomes less effective when local utility electricity price is reduced. Once Scenario A is established in the electricity trading arrangements, householders do not get any extra economic benefit for Microgrid Scenario B. The economic balance is reached when distributed individual CHP penetration gets limited by distribution system. At this point, Microgrid economics is in favour of Microgrid integration. Scenario C is more environmental oriented unless the 50% subsidy of DTI for PV installation is available or installation cost is reduced by 30%. This UK-centric Microgrid economics is applicable to developing countries with poor electricity supply system.

9.6.5   Future developments of Microgrid economics

The economic analysis of DGCG indicates that the electricity demand in a Microgrid can be supplied by micro-CHP with sufficient PV array and battery backup, leading to energy service independent of utility grid. The replacement of household central gas heating systems by micro-CHPs is economically favourable. Hence, the development of infrastructure for a mass market of micro-CHPs would be economically beneficial, leading to enhanced public profile of these technologies. It calls for the development of energy service companies with expertise in energy saving measures and supply demand contracting with renewables. It would be favourable for the establishment of new open market structure for Microgrids with regulation support. New technologies always face economic and regulatory barriers before taking off. Microgrid needs supportive public environment as well as policy and commercial supports for its growth, which is highly suitable to energy policy as per global climate requirement.

Besides, its rapidly expanding market needs to speed up necessary regulatory changes. The new technology should incorporate IT enabled services in a commercial scale for better real-time control of Microgrids. Microgrid's active demand management as per electricity price signals can reduce the peak demands on the network, leading to large-scale energy savings inculcating a culture favourable to household energy efficiency. A Microgrid with the combination of PV, micro-CHP and small storage battery is quite independent of utility grid and hence can be installed at remote communities without any need for transmission cables. There will be no distribution cost outside

Microgrid. Microgrid is also socially beneficial to remove unsightly power cables, enabling better natural aesthetics. This technology can reduce the maximum demand of current central generation system leading to large savings in operation and long-term investments. Microgrids can further incorporate other energy saving measures reducing domestic and office energy demand and GHG emissions from buildings as per Kyoto Protocols. The full potential of Microgrids can only be exploited if necessary changes are effected in the electricity market and regulatory structure. For further details, please refer to Appendix F.

9.7   Conclusion

In the perspectives of electricity industries and government regulatory bodies, it has been examined that Microgrids can integrate small-scale generators in the proximity of consumer premises with several advantages. Microgrid can coexist with utility grid as a citizen having necessary power import and export with the local distribution networks. Microgrids can exert integral control with local distribution networks for power balance, frequency and voltage control, supply reliability and power quality. The technical and regulatory issues associated with the utility integration of Microgrids need to addressed as per the laws of the lands for different countries. The diversity of generation should be appropriately employed for satisfying energy demand permitting stand-alone operation of Microgrids whenever necessary. The current judicial mix of Microgrid DERs is micro-CHP, PV and battery storage. A part of the household micro-CHPs can be replaced by fuel cells when these are commercially available as domestic CHP generators. Economic analyses of Microgrids against the background of current regulatory and economic framework for DG indicate that Microgrids can potentially contribute to an appreciable part of energy demand with several economic benefits. The short-term benefits of Microgrids are very much applicable for long-term perspectives with more possibilities. Microgrids can directly facilitate the implementation of other energy saving measures reducing domestic and office energy demands and GHG emissions as per Kyoto Protocols. It can principally be concluded that Microgrids do have the potential for making major contribution to GHG emission reduction from buildings subject to the necessary major changes in the electricity market and regulatory structure. Its main starting point would be the initiation of all the aforesaid changes for greater benefits to the society.

Microgrids, which are viewed as aggregated controllable load units with their own on-site power generation, can therefore easily participate to sell power and ancillary services in the newly restructured open market. It ensures system reliability, power quality and efficiency at comparatively lower costs.

The Federal Energy Regulatory Commission (FERC) has issued the Order 888, a final rule in 1996 in the USA to enforce open access non-discriminatory tariffs in order to facilitate unbundling of wholesale generation and transmission services. Moreover, FERC has also issued the Order 889 for the development of the electronic communication system termed as Open Access Same-Time Information System (OASIS) that has facilitated the development of three main restructuring models namely PoolCo Model, Bilateral Contracts Model and Hybrid Model. Microgrid market participation is discussed in the following sections.

10.2   Restructuring models

In order to alleviate the monopoly of vertically integrated utilities, three major models have been developed for electricity market restructuring, viz. PoolCo Model, Bilateral Contracts Model and Hybrid Model. These models encompass different types of competition in the open market for ensuring better service to the customers.

10.2.1   PoolCo Model

PoolCo Model is defined as a centralised marketplace that clears the market for buyers and sellers. Here the electric power participants submit bids and prices into the pool for the amounts of energy to sell or buy respectively. The market clearing ensures participation of all the suppliers and customers in the market pool. The Independent System Operator (ISO) normally forecasts the demand for the following day and receives bids to satisfy the demand at the lowest cost and prices for the electricity on the basis of the most expensive generator in operation. The main characteristic of this model is to establish independent wholesale power pool served by interconnected transmission systems. This is a centralised clearing market of electricity trading, aimed at introducing competition that forces the distribution utilities to buy power from the PoolCo and generation utilities to sell power to the PoolCo at a market clearing price (MCP) defined by the PoolCo irrespective of the generation cost. The most widely used MCP is 'the price of the highest selected bid' and the final spot market price may exceed MCP to account for ISO's charges for the associated ancillary services as well as overhead costs. The PoolCo does not own any generation or transmission components and centrally dispatches whole power within its service jurisdiction. It controls the maintenance of transmission grid for ensuring efficient operation by charging non-discriminatory fees to generators and distributors to cover its operating costs. In the PoolCo, both sellers and buyers compete for the rights of injecting and drawing power to and from the PoolCo. Sellers fail to inject power if their bid is too high and similarly buyers may not be successful in drawing power if their bid is too low. It ensures implementation of economic dispatch producing a single spot price for electricity, thereby giving participants a clearance for consumption and investment decisions. Thus, the market dynamics drive the spot price to competitive level that is equal to the marginal cost of the most efficient firms. In this market the sellers are paid for their electricity and customers are charged for their consumption on spot.

10.2.2 Bilateral Contracts Model

Bilateral Contracts Model is sometimes called direct access model because here the role of ISO is more limited, thereby allowing buyers and sellers to negotiate directly in the electricity market without entering into pooling arrangement. In order to have successful participation in the model, small customers need to have aggregation to ensure benefits from competition. Thus Microgrid, being an aggregated load with local generations, can significantly participate in the market to harvest the benefits from open competition.

This model establishes non-discriminatory access and pricing rules for transmission and distribution systems assuring guaranteed sale of power over the available transmission and distribution systems. Wholesale suppliers pay to the transmission and distribution companies (DISCOs) for using their transmission and distribution networks. Generation companies (GENCOs) function as suppliers whereas transmission companies (TRANSCOs) act as common carrier to contracted parties leading to mutual benefits and customer choices. DISCOs, on the contrary, function as an aggregator for a large number of retail customers to supply long-term capacity. The contracts take place in terms of price, quantity and location whereas GENCOs inform the generation schedule to the ISO to ensure availability of sufficient resources for finalising transactions to maintain system reliability. In order to alleviate transmission congestion and maintain real-time reliability, suppliers have to provide incremental or decremental energy bids for availing cost based non-discriminatory access to transmission and distribution systems. The system users who lose in the bidding process are left with the alternative of availing the supply to their loads from other power providers or having to modify their load profiles.

10.2.3 Hybrid Model

The Hybrid Model combines various features of the earlier two models. Here the sale and purchase of power through power exchange (PX) are not obligatory and the customers are allowed to sign bilateral contracts with the pool suppliers of their choice. The sellers and buyers can both opt for not signing any bilateral contract availing maximum flexibility to sell and buy power either through pool or by direct bilateral transaction between seller and buyer. The GENCOs opting to compete through pool need to submit competitive bids to the PX. All the bilateral contracts are normally allowed unless the transmission lines are constrained. Loads that are not included in the bilateral contracts are supplied by economic dispatch of GENCOs through bids in the pool. The co-existence of the pool efficiently identifies the energy requirements of the individual customers and thus helps to simplify the energy balance process. The Hybrid Model is significantly flexible in offering either of the marketing options based on prices and services, but it is much costlier because of the co-existence of separate entities of pool and direct dealing in the same system.

10.3   Independent System Operator (ISO)

The functions of ISO in the market participation of Microgrid are discussed in the following sections.

10.3.1   Background

In the cost-of-service regulatory operation, system operators would preserve the system reliability ensuring moment-to-moment matching of generation and load because of the uncertainty in load prediction. Vertically integrated utilities would operate their own system performing economic dispatch of generation managing sales and purchases to and from control areas. In vertically integrated monopoly, utilities established centrally dispatched regional power pools for co-ordinating better planning and operation of generation and transmission amongst their members leading to improved efficiency, maintenance co-ordination and reserve sharing. It results in reduction in costs incurred by the end-users.

There are three types of power pools namely (i) tight power pools, (ii) loose power pools and (iii) affiliate power pools.

 (i) *Tight power pools* – Tight power pools normally function as control areas bounded by interconnection, metering and telemetry. They regulate automatic generation control (AGC) within its boundaries and tie line PXs contributing to interconnection frequency regulation. They also perform unit commitment and power dispatch and transaction scheduling services on second-to-second basis for their members.
 (ii) *Loose power pools* – Loose power pools have a low level of co-ordination in generation and transmission planning and operation unlike tight power pools. These pools provide significant support to the members during emergency conditions. However, they do not provide any control area service.
(iii) *Affiliate power pools* – In affiliate power pools, the aggregated power generation owned by various members is dispatched as a single utility. The members of these pools sign extensive agreements on governing the cost of generation and transmission services.

The transmission access was limited during congestion with significant growth of power suppliers. The members of the vertically integrated utilities would normally prevent other utilities and suppliers from full access to transmission system. Besides, power pools would control access to regional transmission systems making it much more difficult for non-members to use pool members' transmission facilities. Pools also exercised restrictive membership governance leading to closing pool membership to outsiders. These unfair industry practices prevented the growth of open competitive generation market. This was the scenario in the USA that led to the development of ISOs by the intervention of FERC Order 888 insisting transmission owners to provide comparable service to other customers without having own transmission facilities. The owners were enforced to treat their own wholesales and purchases of energy over their own transmission facilities under same transmission tariffs as applicable to others. This was the starting point to isolate transmission ownership from transmission control leading to the development of ISO as encouraged by FERC for implementing open and non-discriminatory access to the transmission facilities.

An ISO is an entity independent of market participants namely generators, transmission owners, DISCOs and end-users whose function is to ensure fair and non-discriminatory access to transmission, distribution and ancillary services to maintain real-time system operation with reliability. The implementation of good ISO ensures independent transmission system operation without any discrimination in power transaction. Besides, FERC Order 889 assures all market participants to obtain relevant transmission information from OASIS.

10.3.2   The role of ISO

The main objective of the ISO is not generation dispatch but matching energy supply to demand to ensure reliable system operation. It can control generation only to the extent necessary to maintain system reliability and optimal transmission efficiency. It would continually assess condition of transmission system and accordingly approve or deny transmission service requests. It is mainly responsible for maintaining real-time transmission system reliability to ensure system integrity, second-to-second supply–demand balance and maintenance of system frequency within acceptable limits. The ISO may schedule power transfers in a constrained transmission system to the extent necessary for reliable operation of the power market. It may also include and control a PX for smooth dispatch of all the generators and setting the energy price on hourly basis as per the highest price bid in the market. As per FERC Order 888, the following six types of ancillary services must be provided: (i) scheduling, control and dispatch services, (ii) reactive supply and voltage control, (iii) regulation and frequency response services, (iv) energy imbalance service, (v) operating reserve, spinning reserve and supplemental reserve services and (vi) transmission constraint mitigation. The FERC Order 888 also implies that

not iterative. The participants submit additional data including individual generation schedules, takeout point for demand, adjustment bids for congestion management and ancillary bids to the PX after the determination of the MCP. It enables the ISO and the PX to know the injection points of individual generating units to the transmission system.

The following terms are used in relation to day-ahead and hour-ahead markets as per the National Education Research and Evaluation Center (NEREC) definitions (www.nerec.com):

(1) *ATC* – The available transfer capability (ATC) is defined as the measure of transfer capability remaining in the physical transmission network for further commercial activity over and above already committed uses.
(2) *TTC* – The total transfer capability (TTC) is defined as the amount of the electric power that can be reliably transferred over interconnected transmission networks subject to certain conditions.
(3) *TRM* – The transmission reliability margin (TRM) is defined as the amount of power transfer capability necessary for ensuring interconnected transmission network security within reasonable range of uncertainties in system conditions.
(4) *CBM* – The capacity benefit margin (CBM) is defined as the amount of power transfer capability reserved by the load serving entities for ensuring access to generation from interconnected systems for meeting up generation reliability requirements.

The above four quantities are related as

$$TTC = ATC + TRM + CBM$$

10.7   Elastic and inelastic markets

An inelastic market cannot provide sufficient signals and incentives to customers for adjusting their demands as per market price. Thus, the customers are not motivated enough to adjust their demands for adapting to market conditions. Therefore, the MCP is determined mainly by the price structure of the suppliers. The power industry has been run for many decades with this inelastic demand or firm load, before the introduction of open access energy markets.

An elastic market, on the contrary, provides sufficient market signals and incentives to the customers encouraging the adjustment of their demands to adapt to market conditions leading to reduction in their overall energy costs.

10.8   Market power

Market power is defined as owning the ability by a single or a group of seller(s) to drive market price over a competitive level leading to control of the total output preventing competitors from relevant market participation for a significant period of time. Thus, market power exerts monopoly preventing open access competition amongst prospective market participants, thereby leading to the deterioration of service quality and reliability, retardation in technological innovation and misallocation of resources. Market power is exercised intentionally when a participant in generation sector owns lion's share of total available generation. It may also be exercised accidentally by transmission constraints that limits the transfer capability in a certain area for maintaining system reliability and forces the customers to purchase power at higher rates from a local supplier. In case of constrained transmission, the remote units are restricted to supply power while local suppliers drive the market prices. Sometimes customers have to pay higher tariff irrespective of energy usage in off-peak periods due to unavailability of hourly metering which facilitates generators to drive up the market prices for their own benefit. The transmission sector also can exercise market power by providing transmission information to some specific affiliated generators and thereby preventing others from participating in the open competition.

Market powers are of two types

(1) Vertical market power
(2) Horizontal market power

10.8.1 Vertical market power

This is the ownership of a single firm or affiliate firms in power generation and market delivery process with control of a bottleneck in the process. The bottleneck comes from transmission lines through which electricity is delivered to intended buyers. The control of the process bottleneck enables the firm and its affiliates to exercise preference over competitive firms, leading to misuse of its control of transmission and distribution facilities.

A successful, well-planned, fully functional ISO can resolve the problems arising out of vertical market power.

10.8.2 Horizontal market power

This is the ability of a dominant firm or group of firms to control generation restricting output and thereby controlling market prices to its own benefit. It arises out of the local control exerted by sufficient concentration in ownership within a defined market area. It might be viewed as the misuse of influence of a particular group in maintaining the supply–demand equilibrium simply by withholding generation that ultimately results in higher market prices.

Herfindahl–Hirschman Index (HHI) gives a quantitative measure of market power. HHI is defined as the weighted sum of market shares of all the participants in the market and quantified as the sum of the squares of market shares of the participants

$$\text{HHI} = \sum_{i=1}^{N} S_i^2$$

where N is the number of participants and S is the market share of the ith participant.

This market power can also be resolved through successful operation of the ISO.

10.9   Stranded costs

Stranded costs are defined as the costs of uneconomic and inefficient commitments or investments that are made by utilities in traditional monopoly regulation but quite unlikely to be recovered by selling electricity in open competitive market. Stranded cost, a terminology of restructuring process, is basically the difference between the costs that are expected to be recovered under the regulation of vertical monopoly and the costs that are recovered in open competitive market.

In vertical monopoly, utilities would recover their costs of business with considerable returns imposing higher rates of electric power on end-users. The change over from vertical monopoly to restructured electricity market is, however, likely to force the inefficient investments to become unrecoverable in the open competitive market. Thus, the recovery of stranded costs remains as a major economic issue in the restructured electricity market.

10.10   Transmission pricing

Transmission grid is the prime issue that dictates market competition and hence transmission pricing becomes a very important factor in an open competitive market. As per the guidelines of FERC, the transmission pricing should meet traditional revenue requirements of transmission owners, reflecting comparability, i.e. equal transmission cost for both the owner and the other participants for the same service. This pricing must also be practical enough to promote fairness and economic efficiency.

In spite of being a small fraction, as compared to power generation and utility operating expenses, transmission cost is quite significant because of the key importance of transmission function in market competition. Besides, transmission systems enhance power generation market efficiencies by providing true price signals that can be used as a criterion for augmenting transmission and generation capacities for accommodating future loads. Additional transmission capacity also relieves transmission constraints thus allowing less-expensive generation technologies to replace the costlier ones, ultimately leading to increased customer savings.

Transmission price is determined by two methods

(1) Contract path method
(2) MW-mile method

10.10.1 *Contract path method*

In the contract path method, the transmission price is determined on the basis of a predefined path of power flow. Transmission pricing in this method is not accurate enough due to the presence of parallel paths for loop flows of power. Thus, the transmission owners might not be fully paid for the actual use of the facilities provided by them. Another shortcoming of this method is pancaking of transmission rates.

The parallel path flows or loop flows are basically the unscheduled power flows that take place on adjoining transmission systems during power transfer in an interconnected power system. Pancaking is defined as power flow crossing the boundary of a contract path of defined transmission ownership. Extra transmission charges are added to the power transaction leading to the increase in overall transmission price. Pancaking effect can be overcome by zonal pricing schemes of the ISO. In this scheme, the ISO-controlled transmission system is divided into zones and the transmission users need to pay on the basis of energy prices for the defined zones. Transmission price is independent of paths between two zones and number of crossings of zones.

10.10.2 *MW-mile method*

The MW-mile method of transmission pricing, used by some ISOs, is based on the distance traversed by the power flow and the amount of power flows in each line. This method can overcome the problems of loop power flows and gives no credit to counter-flows in transmission lines. However, transmission pricing in this method is quite complicated to incorporate all the energy prices of all the transmission lines.

10.11   Congestion management

Congestion is defined as the situation of overloading of transmission lines or transformers. It may take place due to transmission line outages, generator outages, changes in energy demand, uncoordinated power transactions, etc. It leads to prevention of new contracts, infeasibility in existing contracts, additional outages, monopoly of pricing in some regions as well as deterioration and damages of system components. Congestion may be partly prevented by reservations, rights and congestion pricing and corrected by technical controls

10.11.3   Management of inter-zonal and intra-zonal congestions

Transmission network is a major role player in the open access power market. The phase shifters and transformer tap-changers play preventive and corrective roles in congestion management. These controllers help ISO to execute congestion mitigation without rescheduling generation dispatch schedules. Congestion management becomes easier with the implementation of inter-zonal and intra-zonal schemes considering intra-zonal and inter-zonal power flows and their effects on power system. The main objective of the management of these congestions is to minimise the number of adjustments of preferred schedules applying control schemes to minimise inter-zonal interactions considering contingency-constraint limits.

On the contrary to the practices of congestion management, new contracts are identified to redirect power flows on congested lines. Control devices namely phase shifters, transformer tap-changers and FACTS controllers play a vital role in relieving congestion in restructured environment by controlling line flows. Proper co-ordination of phase shifters and transformer tap-changers enhances trading possibilities and feasibility margins leading to the improvement of system performance and thus enables augmentation of more contracts. The ISO need not go for re-dispatch of preferred schedules for congestion management. The ISO can implement more efficient congestion management by (i) considering contingency limits during congestion mitigation, (ii) minimising number of adjustments and (iii) eliminating interactions between inter-zonal sub-problems, intra-zonal sub-problems and cross-border intra-zonal sub-problems.

The ISO performs contingency analysis after receiving preferred schedules from PX and identifying the worst contingency for modelling congestion management. Then the ISO checks the intra-zonal and inter-zonal congestion to minimise total congestion cost irrespective of scheduling co-ordinators' (SCs') preferred schedules. SCs co-ordinate with ISO on behalf of aggregators, retailers and customers for hourly distribution schedule, balanced schedule of generation to be injected into transmission grid and power to be withdrawn from the grid. The ISO uses incremental/decremental bids to relieve congestion. Inter-zonal congestion is more frequent than the intra-zonal one and hence first the inter-zonal contingency is solved followed by intra-zonal ones. All the inter-zonal congestions are checked by ISO one by one and it tries to solve any congestion by the actions of control devices avoiding any change in preferred schedules. Rescheduling of preferred schedules is the last resort for ISO to tackle transmission congestions. In case of detection of no congestion, submitted preferred schedules of PX and SCs are accepted as final real-time schedules.

10.12   Role of Microgrid in power market competition

10.12.1   Retail wheeling

The restructuring of electrical power industry has potential impact on the corporate structure of utilities, pricing of energy products and services, quality of service delivery to the customers and more so, on the selection of energy generation technologies. The ISO interface in the wholesale market at the transmission level is well defined and reasonably accepted by the industry. Now the restructuring needs to extend its competitive market in the level of retailers for the development of open competitive retail market with direct access of the customers. In this market the consumers will be at liberty to choose their energy providers. The introduction of Microgrid needs to further incorporate technological and institutional changes for this retail wheelling.

The new paradigm for marketing of the transmission, distribution and consumption of electricity is retail wheeling with a target to lowering the energy costs. By virtue of retail wheeling, electric utilities can sell energy to remotely located customers. Customers can also purchase energy from remotely located utilities. In order to enhance cost efficiency, transmission and distribution tariffs of intermediaries need to be avoided by utilising DER technologies as Microgrids. Retail wheeling enables consumers to purchase cheaper electricity directly from providers without the involvement of local intermediaries. Hence, the participation of Microgrids in open competitive retail market is likely to be beneficial from customers' point of view. Microgrids can utilise the retail wheeling opportunities to sell power as well as various ancillary services to consumers through open retail market.

The development of retail competition to open services in the revenue cycle should start with metering and billing. Proponents of retail competition should have adequate information regarding customer profile. The retail competition might have to undergo considerable changes to accommodate the new technologies used in distribution and operational paradigm of the Microgrids. Appropriate commercial and regulatory framework should also be created for clearly defining the criteria for participation of Microgrids in the open retail power market.

The establishment of retail wheeling for Microgrids will take place only if the distribution level customers can be encouraged to purchase power and ancillary services from Microgrids. Besides, the retail market must be capable of injecting excess power from Microgrids to the main utility grid through open competition. This retail wheeling of excess energy is dependent on the operating conditions of the sub-station transformers. If the injection of this extra energy to the main utility is prohibited, then this energy will become surplus resulting in imposition of transmission and distribution (T&D) costs and possible congestion penalties on both sellers and purchasers. In order to sell extra energy to other distribution systems, the Microgrids must sell it through the retail market via the concerned transmission system owner/operator in the locality.

10.12.2 Ancillary services

Microgrids can provide potential ancillary services to the power system for maintaining its voltage/frequency profile, stability and reliability. The capability of DERs to quickly ramp up and shut down their generation would definitely enable Microgrids to take the advantage of short-term selling opportunities with a choice of spinning reserve. By virtue of power electronic interfaces (PEIs), Microgrid generators are fully equipped to supply reactive powers for distribution system voltage support. These PEIs can be suitably configured to provide the necessary reactive power for ancillary services. Microgrids are quite capable of selling, through open retail market, the ancillary services for providing support for the stability of system frequency maintaining voltage profiles which are susceptible to load fluctuations and other contingencies in distribution systems.

Open wholesale markets for ancillary services are already available in the transmission level as operated and controlled by the ISOs. The ancillary services provided by Microgrids are quite comprehensive and useful. But the main bottleneck arises from the lack of established open retail market for these ancillary services in distribution level. It leads to adverse effects in retail market competition. There is no established retail market mechanism for equitable treatment of the ancillary services of Microgrid DERs in comparison to those provided by distribution systems. Hence, it creates discrepancy in Microgrid ancillary services irrespective of their cost-effectiveness. Thus with the establishment of power retail market, it is significantly important to establish ancillary services market enabling Microgrids to participate in this open ancillary services market.

In vertically integrated monopoly, ancillary services are provided by central generators that result in significant system losses while delivering these services, thereby increasing the costs of power transmission and distribution. Since the Microgrid DERs are located close to the customers, the full range of ancillary services provided by Microgrids can be purchased at cheaper rates by customers in addition to electrical energy from Microgrid DER generators. It results in drastic reduction of transmission and distribution losses.

The Microgrid DERs would be able to compete fully for ancillary services in open market only when ancillary services would be from energy supply. Therefore, strong policy incentives and effectively open market mechanisms for ancillary services are the prerequisites for making the fullest use of DER potentials in Microgrid paradigm.

10.12.3   Role of aggregators

The direct participation of DERs in retail market might not be feasible if the ISO faces problems in dealing with a large number of individual resources. This would ultimately call for DER integration leading to the development of Microgrids. Integration of large number of DERs, however, would necessitate the installation of extensive communication facilities for handling huge amount of information. In this matter, aggregators can provide substantial help to the ISO in managing DER units, thereby helping to bridge the gaps of control and management issues for both the ISO and the DERs. An aggregator can become the single point of contact for ISO in handling a large number of DERs for a reasonable amount of capacity like the ISO interface with generating resources.

Irrespective of different functions and characteristics, there is extensive interaction between wholesale and retail markets. Aggregators play an indispensable role in a complex power market environment with a large number of participants for the normal operation of the market. Aggregators usually trade electric energy as per its collected supply–demand information. The role of aggregators in the retail market is significantly much more than that in wholesale market. It is because of the presence of large number of participants in the retail market unlike the wholesale market. Besides, most of the DER investors prefer to perform the trade and management of their generated energy engaging a third party with good experience in power marketing and reasonable knowledge of DER functions. The aggregation of DER generation is mainly focused on (i) energy supply of DERs, (ii) energy demand for DERs and (iii) supply of ancillary services of DERs.

The distribution network operator (DNO) verifies and ensures distribution system operation reliability. Aggregators greatly reduce the workload burdens of both ISO and local DNO particularly when the number of retail market participants is very high. Retail market trading for DER energy is depicted in Figure 10.1.

Since the major advantage of DER generation is elimination of transmission and distribution costs, aggregators take care of the local distribution systems. In case of necessity of wheeling surplus energy to other regions,

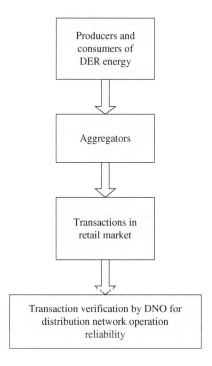

Figure 10.1   DER participation in retail market power trading

Appendix A
Modelling and performance analysis of microturbine in stand-alone and grid-connected modes

A.1 Model description

This section discusses the modelling and simulation of a microturbine–generator (MTG) system consisting of a microturbine (MT) coupled with a synchronous generator. The model is then used to perform the load-following analysis of the MTG system in both stand-alone and grid-connected modes. Simulation is done in MATLAB Simulink platform.

The MTG is analysed for slow dynamic performance of the system and not for transient behaviours. Therefore modelling is based on the following assumptions:

(1) System operation is under normal operating conditions. Start-up, shut-down and fast dynamics (faults, loss of power, etc.) are not included.
(2) The MT's electromechanical behaviour is of main interest. The recuperator is not included in the model as it is only a heat exchanger to raise engine efficiency. Also, due to the recuperator's very slow response time, it has little influence on the timescale of dynamic simulations.
(3) The temperature and acceleration controls have been omitted in the turbine model as they have no impact on the normal operating conditions. Temperature control acts as an upper output power limit. At normal operating conditions, the turbine temperature remains steady, and hence, it can be omitted from the model. Acceleration control is used primarily during turbine start-up to limit the rate of the rotor acceleration prior to reaching operating speed. If the operating speed of the system is closer to its rated speed, the acceleration control could be eliminated in the modelling
(4) Governor model has been omitted as the MT does not use any governor.

The simplified block diagram for MT model is shown in Figure A.1.
The main emphasis is on active power control; therefore, the entire control system is simplified as an active power proportional-integral (PI) control

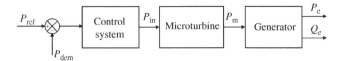

Figure A.1   Microturbine model

function. The controlled active power is applied to the turbine. Active power control is represented as a conventional PI controller as illustrated in Figure A.2.

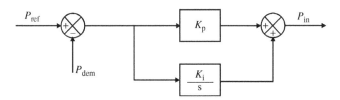

Figure A.2   Controller model

The controller model variables are as follows: P_{in}, active power control variable applied to the input of MT; P_{dem}, actual load demand; P_{ref}, preset power reference; K_p, proportional gain of PI controller; K_i, integral gain of PI controller.

Standard GAST turbine model shown in Figure A.3 is used for simulation. The advantages of GAST model are that it is simple following typical

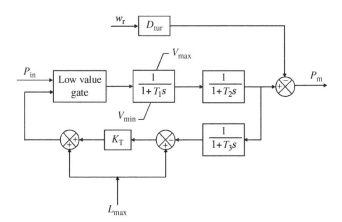

Figure A.3   Turbine model

guidelines and is a WSCC (Western System Co-ordinating Council) model that can be directly used in specific commercial simulation programmes.

The main utility grid is represented by an 11-kV distribution network model using a simple R-L equivalent source of short-circuit level 500 kVA with a load of 5 kW. Grid is integrated into the MT via a 200 kVA, three-phase, 60 Hz, 11/0.440 kV and Y–Δ transformer. The alternator coupled to the MT is modelled as a standard MATLAB Simulink synchronous machine block. The Interconnection between the main grid and the MTG is shown in Figure A.4(a). The MTG is capable of supplying its own loads as well as operating in synchronism with the grid. The MTG can be connected to or disconnected from grid by closing or opening the circuit breaker (CB) at the point of common coupling (PCC). Figure A.4(b) shows the MATLAB model of the MTG system.

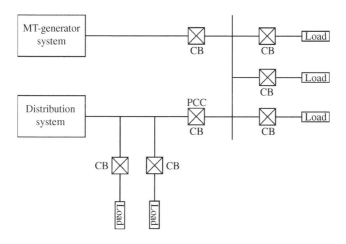

Figure A.4   (a) System configuration block diagram

A.2   Model parameters

The parameters used for simulation of the MT, the alternator and the grid are based on previously reported works and are illustrated in Tables A.1–A.3 respectively.

A.3   Case studies

Following cases have been simulated in MATLAB Simulink environment. Total simulation time for each case is 300 seconds for stand-alone mode and 500 seconds for grid-connected mode. The output powers

A.3.1   Stand-alone mode

A.3.1.1   Case 1

In this case, the stand-alone MTG system is initially running with a load of 30 kW (0.2 p.u.) applied to the generator bus up to $t = 150$ seconds. Another step load of 90 kW (0.6 p.u.) is applied at $t = 150$ seconds. The load on the MTG system is shown in Figure A.5.

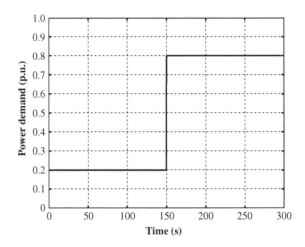

Figure A.5   Load on MTG system

Figure A.6(a) shows the mechanical power output of MT. It is observed that MT power output takes about 90 seconds to match the load demand. MTG speed plotted in Figure A.6(b) shows that MTG system takes almost the same time to reach the new steady-state speed at the new load.

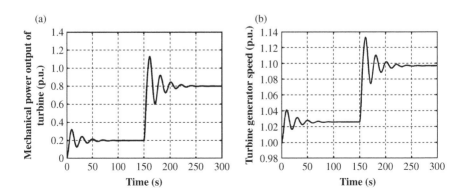

Figure A.6   (a) MT mechanical power; (b) MTG speed

The electrical power output of the generator is shown in Figure A.7. It is seen to closely follow the step change in load demand.

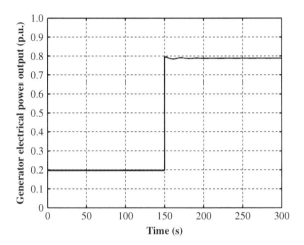

Figure A.7   Generator electrical power

A.3.1.2   Case 2

In this case, a speed control has been incorporated in the stand-alone MTG system to maintain the speed constant at 1 p.u. The MTG is running initially at no load. At time $t = 50$ seconds a load of 0.2 p.u. is applied and at $t = 200$ seconds another load of 0.6 p.u. is applied. The mechanical power output of the MT shown in Figure A.8 indicates that the MT follows the load demand with a time lag of approximately 50 seconds.

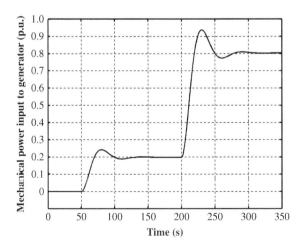

Figure A.8   MT mechanical power

The generator power output shown in Figure A.9 indicates that it closely follows the load as in Case 1. The plot of MTG speed shown in Figure A.10 indicates that speed reaches 1 p.u. and is maintained at that level at the new load.

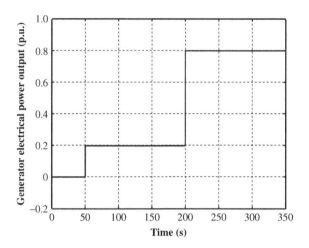

Figure A.9   Generator electrical power

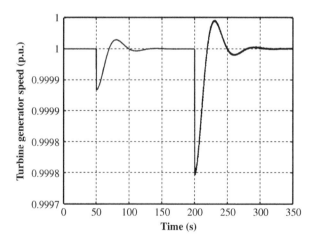

Figure A.10   MTG speed

A.3.2   Grid-connected mode

A.3.2.1   Case 1

In this mode, the MTG system is connected to the utility grid. Initially, both MTG system and grid are running separately at no load. At $t = 5$ seconds, loads of 0.2 p.u. and 160 kW (1.07 p.u.) are applied separately to the MTG and the grid, respectively. At $t = 125$ seconds another load of 0.6 p.u. is applied to MTG. At $t = 250$ seconds, the MTG is interconnected with the grid and at $t = 375$ seconds it is again disconnected from the grid. The MT mechanical power output and the generator electrical power output are shown in Figures A.11 and A.12, respectively.

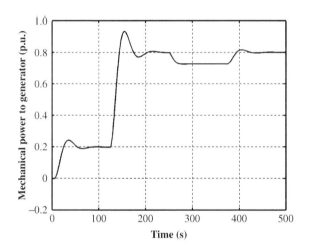

Figure A.11   MT mechanical power

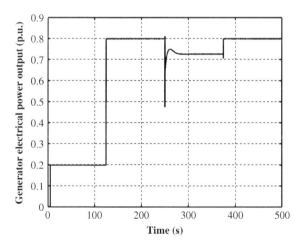

Figure A.12   Generator electrical power

The responses show that load on MTG reduces to some extent due to grid support when it is grid-connected from $t = 250$ to 375 seconds. When stand-alone, the MTG is taking up its entire load of 0.8 p.u. Figure A.13 shows that the

The plot of MT mechanical power in Figure A.16 indicates that it effectively follows the power demand. Then it shares a portion of grid power when connected to the grid at $t = 100$ seconds. MT power again increases when the power demand is increased. It has been observed that the MT supplies 0.7623 p.u., whereas the remaining portion is shared by the grid. Figure A.17 shows that the generator electrical power is found to follow the same pattern as MT mechanical power. Grid power plot of Figure A.18 illustrates that grid power decreases when it is connected with the MTG system as they share each other's demands. However, grid shares a portion of MTG demand as the load on the MTG system is increased at $t = 150$ seconds.

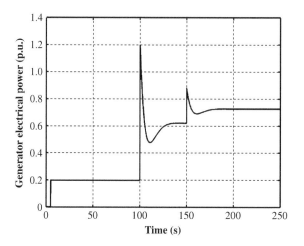

Figure A.17   Generator electrical power

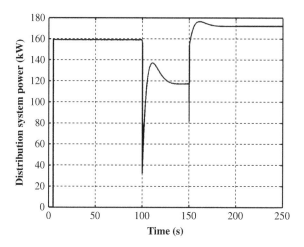

Figure A.18   Grid power

A.3.2.3   Case 3

In this case, the grid load is varied when it is connected with the MTG system. Initially, both the MTG system and the grid are running separately at no load. At $t = 5$ seconds, loads of 0.5 p.u. and 160 kW (1.07 p.u.) are applied separately to the MTG and the grid, respectively. At $t = 100$ seconds the MTG system is inter-connected with the grid and at $t = 150$ seconds another load of 40 kW (0.2667 p.u.) is applied to the grid. The MT mechanical power output, the generator electrical power output and grid power are shown in Figures A.19–A.21, respectively.

The plot of MT mechanical power in Figure A.19 shows that it effectively follows the power demand. Then it shares a portion of grid power when

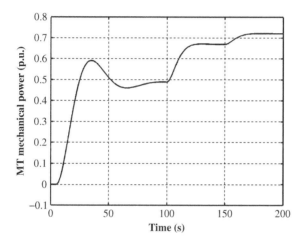

Figure A.19   MT mechanical power

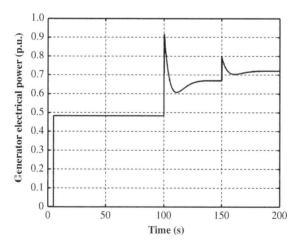

Figure A.20   Generator electrical power

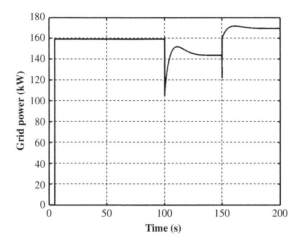

Figure A.21   Grid power

connected to the grid at $t = 100$ seconds. The power again increases when the power demand is increased on the grid. It has been observed that the MT supplies 0.7213 p.u. as it shares a portion of connected load to the grid. The generator electrical power is found to follow the MT mechanical power as illustrated in Figure A.20. The grid power shown in Figure A.21 illustrates that its power decreases when it is connected with the MTG system. However, it shares a portion of MTG demand when load on the MTG system is increased at $t = 150$ seconds.

A.3.3   MTG performance

For performance analysis, the MTG is simulated under different load conditions. Table A.4 shows the MTG operating parameters – viz. generator voltage, current, turbine torque and speed – under different loading conditions as obtained from simulation without speed control and stand-alone mode of the set.

Table A.4   Generator load versus operating parameters

Load (kW)	Voltage (V)	Current (A)	Turbine torque (p.u.)	MTG speed (p.u.)
15	440.05	19.68	0.0998	1.01
30	440.05	39.36	0.1998	1.02
45	440.05	59.04	0.2998	1.03
60	439.99	78.73	0.3997	1.05
75	440.01	98.41	0.4996	1.06
90	439.97	118.10	0.6004	1.07
105	439.98	137.78	0.6998	1.08
120	439.99	157.46	0.8003	1.09
135	440.03	177.14	0.8996	1.10
150	440.01	196.82	0.9997	1.17

Dynamic modelling and performance analysis of a DFIG wind energy conversion system

List of symbols

V_{ds}, V_{qs}, V_{dr}, V_{qr}	d–q axis machine voltages
i_{ds}, i_{qs}, i_{dr}, i_{qr}	d–q axis machine currents
φ	flux linkage
ω_b	base electrical frequency
ω	angular velocity of reference frame
ω_r	angular frequency of rotor
T_{em}	electromagnetic torque
T_{mech}	externally applied mechanical torque
T_{damp}	damping torque
H	inertia constant
r	resistance
x	reactance
d	suffix for direct axis
q	suffix for quadrature axis
s, r	suffices for stator and rotor, respectively
l, m	suffices for leakage and magnetising

B.1 Model description

This section describes an integrated dynamic model of a 750 W variable speed doubly fed induction generator (DFIG)-based wind energy conversion system (WECS). Separate mathematical models are developed for wind flow, rotor, gear and the DFIG using MATLAB Simulink. The WECS model is also validated using realistic data. The model is helpful in choosing an appropriate WECS for any given wind regime.

A typical WECS is shown in Figure B.1. Intermediate output of different models and the overall electrical power output of the proposed WECS are presented in this appendix.

Considering the actual and real configuration of wind farms (electronic components, generator, protection systems, etc.), the developed solution must

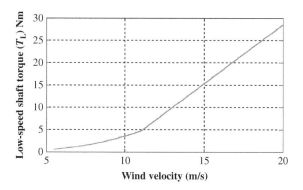

Figure B.4   Low-speed shaft torque, T_L

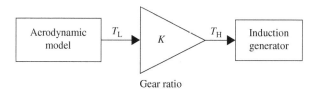

Figure B.5   Gear model

the losses in the gearbox are zero, thus gear transmits ideally from low speed
to high speed:

$$T_H = \frac{T_L}{\eta_{gear}}, \text{ where } \eta_{gear} = \text{gear ratio} \tag{B.4}$$

Figure B.6 shows a sample output of the model.

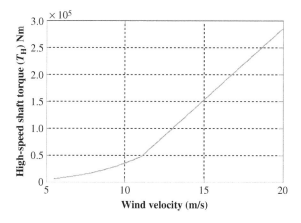

Figure B.6   High-speed shaft torque

B.1.3 DFIG model

The wound rotor-type doubly fed induction generator (DFIG) model is selected for study because it is robust and easy to control from both rotor and stator side. While modelling a DFIG under different wind speeds, it is important to consider that this kind of wound rotor machine has to be fed from both stator and rotor sides. Normally, the stator is directly connected to the grid and the rotor side is interfaced through a variable-frequency power converter. In order to cover a wide operation range from subsynchronous to supersynchronous speeds, the power converter placed on the rotor side has to operate with power flowing in both directions.

B.1.3.1 d–q Model of DFIG

The d–q axis representation of induction generator is used for simulation, taking flux linkage as basic variable. It is based on fifth-order two axis representations commonly known as the 'Park model'. Here an equivalent two-phase machine represents a three-phase machine, where d^s–q^s correspond to *stator direct and quadrature axes*, and d^r–q^r correspond to *rotor direct and quadrature axes*. A synchronously rotating d–q reference frame is used with the direct d-axis oriented along the stator flux position. In this way, decoupled control between the electrical torque and the rotor excitation current is obtained. The reference frame is rotating with the same speed as the stator voltage. While modelling the DFIG, the generator convention is used, indicating that the currents are outputs and that power has a negative sign when fed into the grid.

B.1.3.2 Axes transformation

The d–q model requires that all the three-phase variables have to be transformed into the two-phase synchronously rotating frame. A symmetrical three-phase induction machine with stationary axes as-bs-cs separated by $2\pi/3$-angle is considered. Here the three-phase stationary reference frames (d^s–q^s) variables are transformed into synchronously rotating reference frame (d^e–q^e). Assume that the d^s–q^s axes are oriented at an angle of θ. The voltages V_{ds}^s and V_{qs}^s can be resolved into as-bs-cs components and can be represented in the matrix form as follows:

$$
\begin{bmatrix} V_{as} \\ V_{bs} \\ V_{cs} \end{bmatrix} = \begin{bmatrix} \cos\theta & \sin\theta & 1 \\ \cos(\theta-120) & \sin(\theta-120) & 1 \\ \cos(\theta+120) & \sin(\theta+120) & 1 \end{bmatrix} \begin{bmatrix} V_{qs}^s \\ V_{ds}^s \\ V_{0s}^s \end{bmatrix} \tag{B.5}
$$

$$
\begin{bmatrix} V_{qs}^s \\ V_{ds}^s \\ V_{0s}^s \end{bmatrix} = \left(\frac{2}{3}\right) \begin{bmatrix} \cos\theta & \cos(\theta-120) & \cos(\theta+120) \\ \sin\theta & \sin(\theta-120) & \sin(\theta+120) \\ 0.5 & 0.5 & 0.5 \end{bmatrix} \begin{bmatrix} V_{as} \\ V_{bs} \\ V_{cs} \end{bmatrix} \tag{B.6}
$$

where V_{0s}^s is added as the zero sequence component, which may not be present. Equation (B.6) represents transformation of three-phase quantities into two-phase (d–q) quantities. The current and flux linkages can be transformed by similar equations.

The main variables of the machine in rotating frame are flux linkages φ_{qs}, φ_{ds}, φ'_{qr} and φ'_{dr} in state space form, which are given as follows:

$$\varphi_{qs} = \omega_b \int \left(\left(V_{qs} - \left(\frac{\omega_r}{\omega_b \varphi_{ds}} \right) + \frac{(r_s/x_{ls})}{\varphi_{mq} - \varphi_{qs}} \right) \right) \tag{B.7}$$

$$\varphi_{ds} = \omega_b \int \left(\left(V_{ds} - \left(\frac{\omega_r}{\omega_b \varphi_{qs}} \right) + \frac{(r_s/x_{ls})}{\varphi_{md} - \varphi_{ds}} \right) \right) \tag{B.8}$$

$$\varphi'_{qr} = \omega_b \int \frac{r'r/x_{lr}}{\varphi_{mq} - \varphi'_{qr}} \tag{B.9}$$

$$\varphi'_{dr} = \omega_b \int \frac{r'r/x_{lr}}{\varphi_{md} - \varphi'_{dr}} \tag{B.10}$$

Currents can be calculated by substituting conditions $\omega = \omega_r$ and $V_{qr} = V_{dr} = 0$ in (B.7)–(B.10):

$$i_{qs} = \frac{(\varphi_{qs} - \varphi_{mq})}{x_{ls}} \tag{B.11}$$

$$i_{ds} = \frac{(\varphi_{ds} - \varphi_{md})}{x_{ls}} \tag{B.12}$$

$$i'_{qr} = \frac{(\varphi'_{qr} - \varphi_{mq})}{x'_{lr}} \tag{B.13}$$

$$i'_{dr} = \frac{(\varphi'_{dr} - \varphi_{md})}{x'_{lr}} \tag{B.14}$$

Solving (B.11)–(B.14), φ_{mq} and φ_{md} are obtained as follows:

$$\varphi_{mq} = x_m \left(\frac{\varphi_{qs}}{x_{ls}} + \frac{\varphi'_{qr}}{x'_{lr}} \right) \tag{B.15}$$

$$\varphi_{md} = x_m \left(\frac{\varphi_{ds}}{x_{ls}} + \frac{\varphi'_{dr}}{x'_{lr}} \right) \tag{B.16}$$

where $x_m = 1/(1/x_m + 1/x_{ls} + 1/x_{lr})$.

Electromagnetic torque of the induction generator, expressed in terms of d–q axis flux linkages and currents, is given by

$$T_{\text{em}} = \left(\frac{3p}{4\omega_{\text{b}}}\right)(\varphi_{\text{ds}}i_{\text{qs}} - \varphi_{\text{qs}}i_{\text{ds}}) \text{ Nm} \tag{B.17}$$

The equation that governs the motion of rotor is obtained by equating the inertia torque to the accelerating torque:

$$J\left(\frac{d\omega_{\text{rm}}}{dt}\right) = T_{\text{em}} + T_{\text{mech}} - T_{\text{damp}} \text{ Nm} \tag{B.18}$$

Expressed in per unit values, (B.18) becomes

$$\frac{2Hd(\omega_{\text{r}}/\omega_{\text{b}})}{dt} = T_{\text{em}} + T_{\text{mech}} - T_{\text{damp}} \text{ Nm} \tag{B.19}$$

In (B.17) and (B.19), the flux linkages are two-phase d–q axes rotating reference frame. To obtain the three-phase output, the two-phase d–q axes quantities have to be transformed into three-phase stationary reference frame. This is achieved with the help of following steps.

(1)   Zero-n conversion
In case of an isolated neutral system, the phase voltages are obtained using zero sequence voltage. The transformation is represented in the matrix form as

$$\begin{bmatrix} V_{an} \\ V_{bn} \\ V_{cn} \end{bmatrix} = \begin{bmatrix} 2/3 & -1/3 & -1/3 \\ -1/3 & 2/3 & -1/3 \\ -1/3 & -1/3 & 2/3 \end{bmatrix} \begin{bmatrix} V_{a0} \\ V_{b0} \\ V_{c0} \end{bmatrix} \tag{B.20}$$

(2)   Unit vector calculation
Unit vectors $\cos\theta_{\text{e}}$ and $\sin\theta_{\text{e}}$ are used in vector rotation, 'abc-dq conversion' and 'dq-abc conversion'. The angle θ_{e} is calculated directly by integrating the frequency of the input three-phase voltages ω_{e}. The unit vectors are obtained simply by taking the sine and cosine of θ_{e}. This helps to observe the rotor position in d–q model.

(3)   abc-dq and dq-abc Conversion
The two-phase voltages are obtained using abc-dq conversion and are given by

$$\begin{bmatrix} V^{\text{s}}_{qs} \\ V^{\text{s}}_{ds} \end{bmatrix} = \begin{bmatrix} 1 & 0 & 0 \\ 1 & -1/\sqrt{3} & 1/\sqrt{3} \end{bmatrix} \begin{bmatrix} V_{an} \\ V_{bn} \\ V_{cn} \end{bmatrix} \tag{B.21}$$

The three-phase currents are obtained using dq-abc conversion using (B.21) and are given by

$$i^{\text{s}}_{qs} = V_{qs}\cos\theta_{\text{e}} + V_{ds}\sin\theta_{\text{e}} \tag{B.22}$$

$$i^{\text{s}}_{ds} = -V_{qs}\sin\theta_{\text{e}} + V_{ds}\cos\theta_{\text{e}} \tag{B.23}$$

Appendix C
Software simulation of PEM fuel cell system for dynamic performance analysis

C.1 PEMFC power generation system

This section describes two dynamic models of a proton exchange membrane fuel cell (PEMFC) for dynamic performance analysis. The first model uses a proportional-integral (PI) controller to control the fuel flow to the reformer of the fuel cell system and the second model uses a fuzzy logic controller (FLC). The dynamic behaviour of the fuel cell system with a step load change is studied and the outputs are compared for both models.

The PEMFC power generation system has three main parts

(1) Fuel processor
(2) Power section
(3) Power-conditioning unit

The fuel processor converts the fuel into hydrogen and by-product gases. The power section generates electricity using a number of fuel cells. The power-conditioning unit (PCU) converts the generated DC power into output AC power with current, voltage and frequency control to meet the demand as per requirement.

C.2 Dynamic model of PEMFC

The dynamic model of PEMFC is shown in Figure C.1.
The following assumptions are made for the modelling:

(1) The Nernst's equation is applicable.
(2) The gases are ideal.
(3) PEMFC is fed with hydrogen and oxygen.
(4) PEMFC temperature is stable.
(5) The electrode channels are small enough so that the pressure drop across them is negligible.
(6) The ratio of pressure between the inside and outside of the electrode channels is large enough to assume choked flow.
(7) Ohmic and activation losses are taken into consideration.

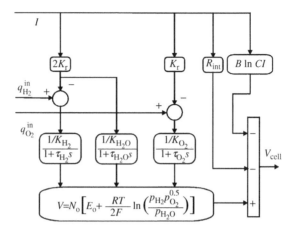

Figure C.1   PEMFC model

The reaction that occurs in a PEMFC is

$$H_2 + \tfrac{1}{2}O_2 \rightarrow H_2O \tag{C.1}$$

The potential difference between the anode and the cathode is calculated using Nernst's equation and Ohm's law:

$$V = N_0 \left[E_o + \frac{RT}{2F} \left(\ln \frac{p_{H_2} p_{O_2}^{0.5}}{p_{H_2O}} \right) \right] - rI - B \ln CI \tag{C.2}$$

Ideal gas law is used to find out the partial pressures of the gases flowing through the electrodes. The same formula is applicable to all the gases. Hence, the expression for hydrogen is given as

$$p_{H_2} V_{an} = n_{H_2} RT \tag{C.3}$$

where p_{H_2} is the partial pressure of hydrogen, T the temperature, V_{an} the volume of anode channel, n_{H_2} the moles of hydrogen in the channel and R the ideal gas constant.

The relationship of gas flow through the valve is proportional to its partial pressure and can be expressed as

$$\frac{q_{H_2}}{p_{H_2}} = \frac{k_{an}}{\sqrt{M_{H_2}}} = k_{H_2} \tag{C.4}$$

and

$$\frac{q_{H_2O}}{p_{H_2O}} = \frac{k_{an}}{\sqrt{M_{H_2O}}} = k_{H_2O} \tag{C.5}$$

where q_{H_2} is the molar flow of hydrogen (kmol/s), q_{H_2O} the molar flow of water (kmol/s), p_{H_2} the partial pressure of hydrogen (atm), p_{H_2O} the partial pressure of water (atm), k_{H_2} the hydrogen valve molar constant (kmol/atm/s), k_{H_2O} the water valve molar constant (kmol/atm/s), k_{an} the anode valve molar constant (kmol-kg$^{0.5}$/atm/s), M_{H_2} the molar mass of hydrogen (kg/kmol) and M_{H_2O} the molar mass of water (kg/kmol).

For hydrogen, the perfect gas equation is used to find the derivative of its partial pressure:

$$\frac{d}{dt}p_{H_2} = \frac{RT}{V_{an}}(q_{H_2}^{in} - q_{H_2}^{out} - q_{H_2}^{r}) \tag{C.6}$$

where R is the real gas constant (1 atm/kmol/K), T the absolute temperature (K), $q_{H_2}^{r}$ the hydrogen flow that reacts (kmol/s), $q_{H_2}^{in}$ the hydrogen input flow (kmol/s) and $q_{H_2}^{out}$ the hydrogen output flow (kmol/s).

The relationship between the hydrogen flow and the stack current can be written as

$$q_{H_2}^{r} = \frac{N_o I}{2F} = 2K_r I \tag{C.7}$$

where N_o is the number of fuel cells in series in the stack, I the stack current (A), F the Faraday's constant (C/kmol) and K_r the modelling constant (kmol/s/A).

Using the equation of hydrogen flow, the derivative of partial pressure can be written in s-domain as

$$p_{H_2} = \frac{1/K_{H_2}}{1 + \tau_{H_2}s}(q_{H_2}^{in} - 2K_r I) \tag{C.8}$$

The partial pressures of water and oxygen can also be calculated using the same expression for partial pressure of hydrogen.

Polarisation curves can be expressed as

$$V_{cell} = V + \eta_{act} + \eta_{ohmic} \tag{C.9}$$

where η_{act} is a function of the oxygen concentration, CO_2 and stack current I (A), and η_{ohmic} is a function of the stack current and the stack internal resistance R_{int} (Ω).

V_{cell} can be rewritten by assuming constant temperature and oxygen concentration:

$$V_{Cell} = V - B\ln(CI) - R_{int}I \tag{C.10}$$

where B and C are constants.

Table C.1   Model Parameters

Parameter	Value
Stack temperature (T)	343 K
Faraday's constant (F)	96,484,600 C/kmol
Universal gas constant (R)	8,314.47 J/kmol/K
No load voltage (E_0)	0.8 V
Number of cells per stack (N_o)	550
Number of stacks (N_{stack})	6
K_r constant $= N_o/(4F)$	1.4251×10^{-6} kmol/s
Utilisation factor (U)	0.8
Hydrogen valve constant (K_{H_2})	4.22×10^{-5} kmol/s/atm
Water valve constant (K_{H_2O})	7.716×10^{-6} kmol/s/atm
Oxygen valve constant (K_{O_2})	2.11×10^{-5} kmol/s/atm
Hydrogen time constant (τ_{H_2})	3.37 s
Water time constant (τ_{H_2O})	18.418 s
Oxygen time constant (τ_{O_2})	6.74 s
Reformer time constant (τ_1)	2 s
Reformer time constant (τ_2)	2 s
Reformer PI gain (C_1)	0.25
Conversion factor (CV)	2
Activation voltage constant (B)	0.04777 A^{-1}
Activation voltage constant (C)	0.0136 V
Internal resistance (R_{int})	0.2778 Ω
External line reactance (X)	0.05 Ω
PI gain constants (C_2)	0.1
PI gain constants (C_3)	10
Methane reference signal ($Q_{methref}$)	0.000015 kmol/s
Hydrogen–oxygen flow ratio (r_{H_O})	1.168
Current delay time constant (T_d)	3 s

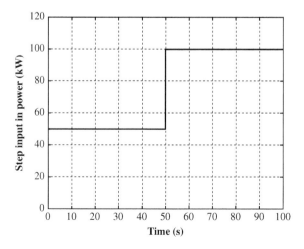

Figure C.4   Step load change

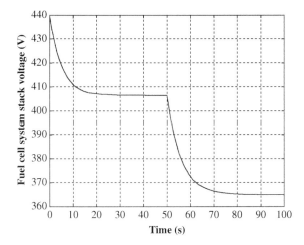

Figure C.5   Stack voltage (PI)

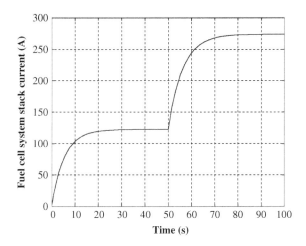

Figure C.6   Stack current (PI)

C.7   Design of fuzzy logic controller

The steps followed for the design of an FLC are as follows:

(1) Fuzzification strategies and the interpretation operator (fuzzifier)
(2) Data base
    (i) Discretisation/normalisation of universe of discourse
    (ii) Fuzzy partition of the input and output spaces

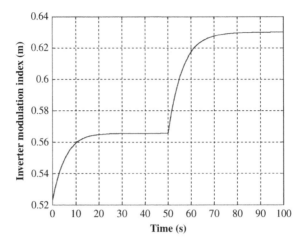

Figure C.7   Inverter modulation index (PI)

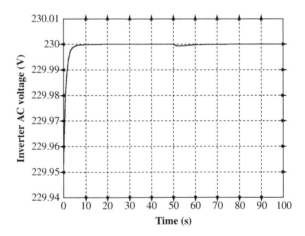

Figure C.8   Inverter AC voltage (PI)

    (iii) Completeness
    (iv) Choice of membership function of a primary fuzzy set
(3) Rule base
    (i) Choice of process state (input) variables and control (output) variables of fuzzy control rules
    (ii) Source and derivation of fuzzy control rules
    (iii) Types of fuzzy control rules
    (iv) Consistency interactivity, completeness of fuzzy control rules
(4) Fuzzy inference mechanism
(5) Defuzzification strategies and the interpretation of defuzzification operators (defuzzifier).

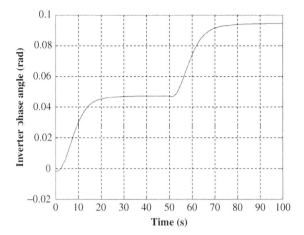

Figure C.9   Inverter phase angle (PI)

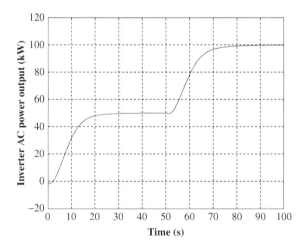

Figure C.10   Inverter AC power output (PI)

The FLC model has two inputs

(1) Error
(2) Rate of change of error

The output of the controller is fed into the input of the reformer. The reformer converts fuel into hydrogen as per the controller output. Each input and output of the FLC has seven membership functions. The fuzzy partitions

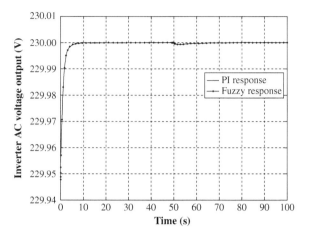

Figure C.14   Inverter AC voltage (FLC)

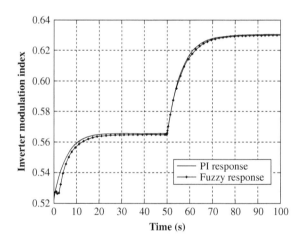

Figure C.15   Inverter modulation index (FLC)

Comparison shows that FLC works better than the conventional PI controller in terms of load-following function, and the performance of the PEMFC system is enhanced.

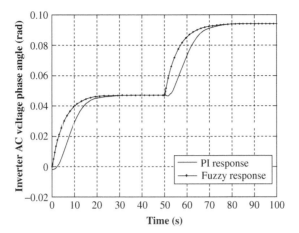

Figure C.16   Inverter phase angle (FLC)

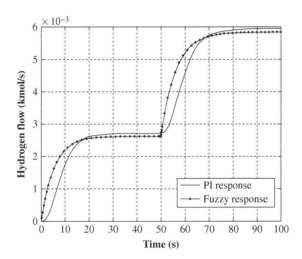

Figure C.17   Hydrogen flow (FLC)

Appendix D
Application of solid-oxide fuel cell in distributed power generation

D.1 SOFC power generation system

This section describes a dynamic model of a 100 kW solid-oxide fuel cell (SOFC) power generation system and its control. This is suitable for studying its performance in distributed generation (DG) systems. The SOFC system is chosen as a distributed energy resource (DER) because of its ability to tolerate relatively impure fuels. It can also be operated at a higher operating temperature. This dynamic model can be used to simulate and analyse the performance of such a system in both stand-alone and integrated modes with other DERs to predict its dynamic behaviour and load-following characteristics. With a strategy developed to control the active power and inverter output AC voltage, it is highly efficient and capable of providing good dynamic behaviour and load-following characteristics while maintaining the load parameters. The proposed model is used to simulate a step change in power demand from the inverter side controller to the SOFC-Inverter system. It uses two proportional-integral (PI) controllers separately with the SOFC system to control the fuel flow in accordance with the power demand and to maintain the bus voltage constant at the set point value.

A power generation fuel cell system has the following three main parts:

(1) Fuel processor
(2) Power section
(3) Power-conditioning unit

The fuel processor converts the fuel into hydrogen and by-product gases. The power section generates electricity using a number of fuel cells. The power-conditioning unit consists of a DC–DC converter that converts the generated DC power into regulated DC output and this regulated DC is converted into AC power by the DC–AC inverter with current, voltage and frequency control to meet the demand as per requirement, as illustrated in Figure D.1.

Considering ohmic losses of the stack and ignoring activation voltage loss, the expression of total stack voltage can be written as

$$V = N_{\mathrm{o}}\left[E_{\mathrm{o}} + \frac{RT}{2F}\left(\ln\frac{p_{\mathrm{H_2}}p_{\mathrm{O_2}}^{0.5}}{p_{\mathrm{H_2O}}}\right)\right] - rI \tag{D.11}$$

where V is the total stack voltage, and rI the ohmic loss of the stack.

The output voltage of the stack is given by the Nernst equation. The ohmic loss of the stack is due to the resistance of the electrodes and the resistance of the flow of oxygen ions through the electrolyte. The activation voltage loss is due to the sluggishness of the reactions at the electrode surfaces. To move the electrons to the electrodes a portion of the voltage is lost in driving the chemical reaction in the fuel cell stack. The total power generated by the fuel cell is

$$P_{\mathrm{FC}} = N_{\mathrm{o}}VI \tag{D.12}$$

D.3   Model of the SOFC power generation system

Figure D.2 shows the SOFC system model used in simulation. The overall reaction of the fuel cell is given by

$$\mathrm{H_2} + \frac{1}{2}\mathrm{O_2} \rightarrow \mathrm{H_2O} \tag{D.13}$$

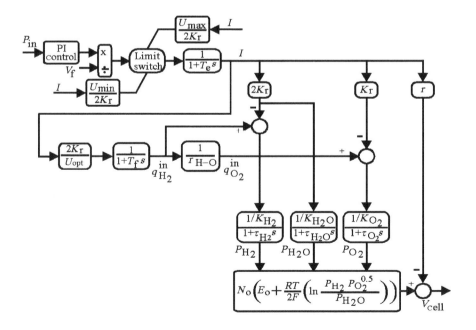

Figure D.2   SOFC system model

Therefore, the stoichiometric ratio of hydrogen and oxygen is 2:1. Excess oxygen is taken in so that hydrogen can react with oxygen fully. The oxygen input flow is controlled by the hydrogen–oxygen ratio r_{H-O}. With the change in the flow of reactants, it takes time to change the parameters of the chemical reaction. Therefore, the chemical response in the fuel processor is slow. This response is represented using a first-order transfer function with a time constant of T_f. The dynamic electrical response is modelled using a first-order transfer function with a time constant of T_e. Electrical response is associated with the speed of the chemical reaction at which charge is restored after it is drained away by load. Power output of the SOFC system is the product of stack current and voltage. The pressure difference between the anode and channels is to be maintained below 4 kPa under normal operating conditions and up to 8 kPa during transient conditions.

D.3.1 Model parameters

The parameters of the system model are given in Table D.1.

Table D.1 Model parameters

Parameters	Values
Absolute temperature (T)	1,273 K
Faraday's constant (F)	96,487,000 C/kmol
Universal gas constant (R)	8341 J/kmol/K
No load voltage (E_o)	1.18 V
Number of cells in series in the fuel cell stack (N_o)	384
Constant $K_r = N_o/4F$	0.99×10^{-6} kmol/s/A
Maximum fuel utilisation (U_{max})	0.9
Minimum fuel utilisation (U_{min})	0.8
Optimal fuel utilisation (U_{opt})	0.85
Valve molar constant for hydrogen (K_{H_2})	8.43×10^{-4} kmol/s/atm
Valve molar constant for water (K_{H_2O})	2.81×10^{-4} kmol/s/atm
Valve molar constant for oxygen (K_{O_2})	2.52×10^{-3} kmol/s/atm
Response time for hydrogen (τ_{H_2})	26.1 s
Response time for water (τ_{H_2O})	78.3 s
Response time for oxygen (τ_{O_2})	2.91 s
Ohmic loss (rI)	0.126 Ω
Fuel system response time (T_f)	5 s
Electrical response time (T_e)	0.0 s
Hydrogen–oxygen ratio (r_{H-O})	1.145
Rated power (P_{rated})	100 kW
Reference power (P_{ref})	100 kW

D.3.2 *Power and voltage control strategy*

Two separate PI controllers are used to control the fuel flow to the system to meet the power demand at the specified inverter output AC voltage. From the SOFC system the voltage is taken as input to the control functions. The controller generates inverter phase angle of the AC voltage and the modulation index.

The AC voltage and active and reactive power outputs of the inverter can be expressed as follows:

$$V_{ac} = mV_{cell}\angle\delta \tag{D.14}$$

$$P_{ac} = \frac{mV_{cell}V_s}{X}\sin\delta \tag{D.15}$$

$$Q_{ac} = \frac{mV_{cell}^2 - mV_{cell}V_s}{X}\sin\delta \tag{D.16}$$

where m is the modulation index, V_{cell} the SOFC DC voltage, V_s the load voltage, δ the phase angle of the AC voltage and X the external line reactance.

Assuming a lossless inverter,

$$P_{ac} = P_{dc} = V_{cell}I \tag{D.17}$$

Now, hydrogen flow can be expressed as

$$q_{H_2} = \frac{2K_r I}{U_{opt}} \tag{D.18}$$

where I is the FC rated current, K_r the constant and U_{opt} the optimum fuel utilisation.

Using (D.18) and assuming δ to be very small, δ can be expressed as

$$\delta = \frac{U_{opt}X}{2K_r mV_s}q_{H_2} \tag{D.19}$$

Equation (D.19) indicates how AC voltage phase angle can be controlled by controlling the hydrogen flow.

With the expressions of AC power and phase angle, it is now possible to control the active power output by the use of hydrogen flow. The power and voltage control strategy is shown in Figure D.3.

Actual inverter AC power and voltage are fed back into the controllers to generate modulation index and phase angle for the AC voltage.

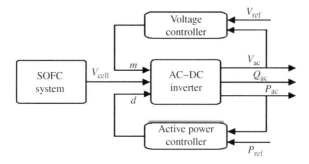

Figure D.3   Power and voltage control strategy

D.4   Case study

To analyse the dynamic behaviour of the SOFC system, a step load from 50 to 100 kW is applied $t = 50$ seconds. The voltage controller is simultaneously set to maintain the inverter output AC voltage at the specified value. The simulation results obtained are shown in Figures D.4–D.14 that illustrate the dynamic behaviour of the system under step load change condition keeping the inverter AC voltage at constant level. The change in stack voltage, change in stack current, change in output power, inverter AC voltage, change in hydrogen flow, change in oxygen flow and the pressure difference between anode and cathode were observed and are represented as obtained from the simulation.

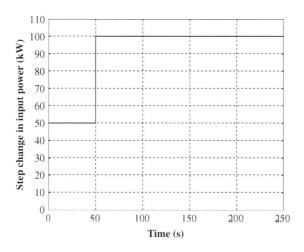

Figure D.4   Step change in power demand

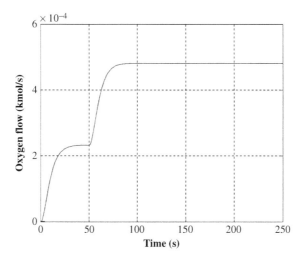

Figure D.9   Oxygen flow

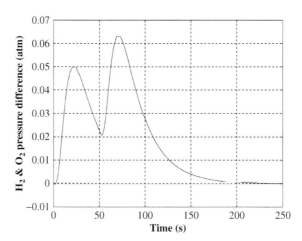

Figure D.10 H_2–O_2 pressure difference

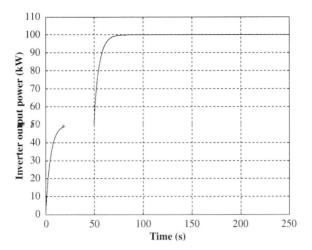

Figure D.11 Inverter AC power output

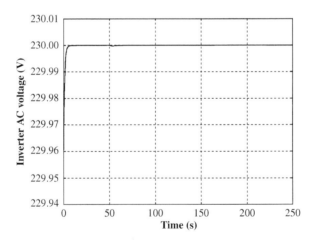

Figure D.12 Inverter AC voltage

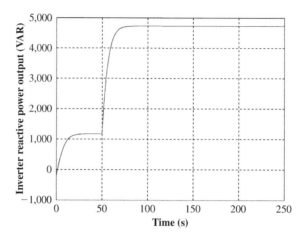

Figure D.13   Inverter reactive power

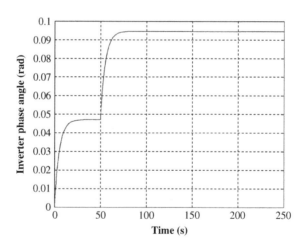

Figure D.14   Inverter AC voltage phase angle

Appendix E
Modelling and performance evaluation of a stand-alone photovoltaic (PV) plant with maximum power point tracking

List of symbols

I_{ph}	photocurrent of the double diode model
I_{s1}, I_{s2}	saturation currents of the diode terms in the double diode model
R_s	series resistance in the double diode model
R_p	parallel resistance in the double diode model
A	diode parameter
V	cell terminal voltage in V
I	cell terminal current in A
E	solar irradiance in W/m^2
T	ambient temperature in K
k	Boltzmann's constant
e	electronic charge
V_{oc}	open-circuit voltage of each cell
P_m	maximum power of each cell
V_{pm}	maximum power voltage of each cell
I_{pm}	maximum power current of each cell
N_s	number of cells in series
N_p	number of cells in parallel
V_{ocar}	open-circuit voltage of the array
P_{mar}	maximum power of the array
V_{pmar}	maximum power voltage of the array
I_{pmar}	maximum power current of the array
R_{in}	internal resistance of each cell
R_{inar}	internal resistance of the array

E.1 Photovoltaic modelling

Photovoltaic (PV) plants consist of inverter-interfaced PV arrays. The inverter keeps the AC output voltage at the specified level irrespective of solar irradiance E (W/m^2) and ambient temperature T (K). The inverters are provided with maximum power point tracking (MPPT) feature that sets the operating

In the model, the PV array has been assumed to be comprising a combination of N_s cells in series and N_p cells in parallel in order to achieve the required voltage, current and power capacity. The terminal DC voltage of the array is maintained at its maximum power voltage V_{pmar} corresponding to the solar irradiance and ambient temperature condition while the value of the ideal DC voltage source is set at the open-circuit voltage V_{ocar} for the array. V_{pmar}, V_{ocar} and the internal resistance R_{inar} for the array have been calculated from the values of the open-circuit voltage (V_{oc}), maximum power voltage (V_{pm}) and maximum power current (I_{pm}) of each cell. V_{oc}, V_{pm} and I_{pm} have been determined by generating the V–I characteristics and the power curve of each cell through the modelling programme according to the double diode model described in (E.1). The flow chart for the modelling programme is shown in Figure E.3.

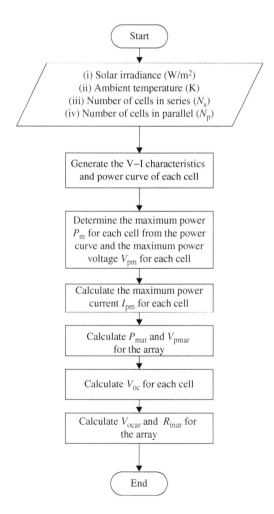

Figure E.3   Flow chart for calculation of PV array equivalent circuit

The detailed mathematical calculations for various parameters are listed in Sections E.3.1–E.3.4.

E.3.1 Calculation of V_{oc}

The open-circuit voltage V_{oc} for a single cell is determined by setting $I = 0$ and $V = V_{oc}$ in the double diode model as shown in (E.9). V_{oc} is the maximum value that terminal voltage V can achieve with full open circuit:

$$V_{oc} = R_{p}[I_{ph} - I_{s1}(e^{(V_{oc}/v_{t})} - 1) - I_{s2}(e^{(V_{oc}/Av_{t})})] \tag{E.9}$$

E.3.2 Calculation of P_m, V_{pm} and I_{pm}

The V–I characteristics and power curve of a cell are generated by varying V from 0 to V_{oc} and by storing the values of V, I and $P = VI$ as separate arrays. The power array is then searched for its maximum value $P_m = P[i]$ and the value of maximum power voltage V_{pm} is obtained by setting $V_{pm} = V[i]$ where i is the co-ordinate corresponding to maximum power P_m. Similarly, maximum power current I_{pm} is set as $I_{pm} = I[i]$.

E.3.3 Calculation of V_{ocar}, P_{mar} and V_{pmar}

For the array the open-circuit voltage, maximum power and maximum power voltage are calculated as

$$V_{ocar} = V_{oc}N_{s} \tag{E.10}$$

$$P_{mar} = P_{m}N_{s}N_{p} \tag{E.11}$$

and

$$V_{pmar} = V_{pm}N_{s} \tag{E.12}$$

respectively. Though R_{inar} can be calculated from V_{oc}, V_{pm} and I_{pm} only, V_{pmar} and P_{mar} have been computed for the purpose of performance evaluation of the array.

E.3.4 Calculation of internal resistance R_{inar}

During calculation of the internal voltage of the array, it has been assumed that the change from V_{oc} to V_{pm} for each cell is linear, i.e. a change from current $I = 0$ to $I = I_{pm}$ has caused the terminal voltage to drop from V_{oc} to V_{pm}. It is also assumed that the DC voltage at the converter terminal must be equal to V_{pm}. Thus, for each cell, the value of the internal resistance can be calculated as $R_{in} = (V_{oc} - V_{pm})/I_{pm}$. For the array, N_{s} series and N_{p} parallel combination of R_{in} gives the value of R_{inar} as follows:

$$R_{inar} = \frac{V_{oc} - V_{pm}}{I_{pm}}\frac{N_{s}}{N_{p}} \tag{E.13}$$

In the calculation, the values of the K coefficients used are as reported in earlier works. The values of K coefficients and those of E, T, N_s and N_p are listed in Table E.1.

Table E.1   Values of input coefficients and parameters

Input coefficients/parameters	Value
K_0	$-5.729e^{-7}$
K_1	-0.1098
K_2	44.5355
K_3	$-1.2640e^4$
K_4	11.8003
K_5	$-7.3174e^3$
K_6	2
K_7	0
K_8	1.47
K_9	$1.6126e^{-3}$
K_{10}	$-4.47e^{-3}$
K_{11}	$2.3034e^6$
K_{12}	$-2.8122e^{-2}$
N_s	$2,000$
N_p	400

E.4   Case studies and results

E.4.1   Case 1

(I) Generation of V–I characteristics and power curves for varying solar irradiances and constant ambient temperature.

Figure E.4(a) and E.4(b) shows the V–I and power characteristics generated by the programme for a fixed ambient temperature of 25 °C ($T = 298$ K) and irradiance values of $E -$ 500, 600, 700, 800, 900 and 1,000 W/m². Table E.2 gives the values of V_{ocar}, V_{pmar}, P_{mar} and R_{inar} for the same condition.

(II) Generation of V–I characteristics and power curves for varying ambient temperature and constant solar irradiance.

Figure E.5(a) and E.5(b) shows the V–I and power characteristics generated by the programme for a fixed solar irradiance of $E = 1,000$ W/m² and ambient temperature values of 5, 15, 25, 35 and 45 °C (i.e., $T = 278, 288, 298, 308$ and 318 K respectively). Table E.3 gives the values of V_{ocar}, V_{pmar}, P_{mar} and R_{inar} for these conditions.

E.4.2   Case 2

The model has been used to develop a stand-alone PV system that delivers AC load through DC–AC inverter interface as shown in Figure E.6. This has been

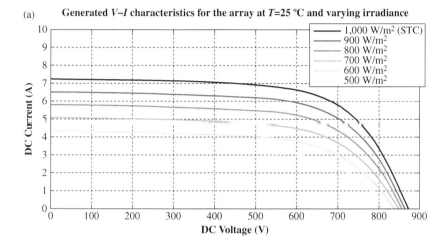

(a) Generated *V–I* characteristics for the array at *T*=25 °C and varying irradiance

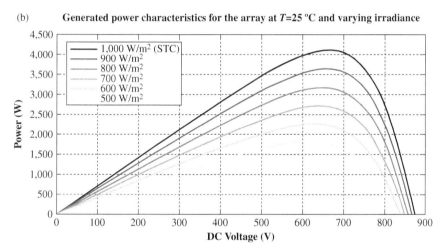

(b) Generated power characteristics for the array at *T*=25 °C and varying irradiance

Figure E.4   *(a) Generated* V–I *characteristics for varying* E; *(b) Generated power characteristics for varying* E

Table E.2   *Values calculated at ambient temperature of 25 °C for different values of solar irradiance*

E (W/m^2)	V_{ocar} (V)	V_{pmar} (V)	P_{mar} (W)	R_{inar} (Ω)
1,000	873.42	660	4,124	34.16
900	866.42	660	3,652	37.31
800	858.47	660	3,181	41.18
700	849.29	630	2,721	50.78
600	838.44	630	2,270	57.85
500	825.22	630	1,821	67.54

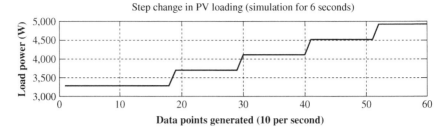

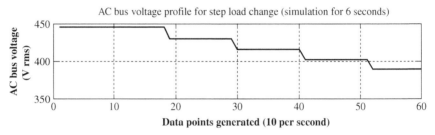

Figure E.7   AC voltage profile for step load change for E = 1,000 W/m² and
T = 298 K

voltage profile with load changes from 3,690 to 4,920 W in five steps over a
period of 6 seconds.

(II) E = 800 W/m² and T = 298 K (25 °C)

Table E.5 shows the values of AC bus voltage for different loadings for the value
of $E = 800$ W/m² and $T = 298$ K (25 °C). It is clear from Table E.5 that AC side
loading on the array that keeps the bus voltage fixed at 415 V is almost 3,160 W,
which corresponds to the maximum power capacity of 3,181 W as obtained
from Table E.2. Figure E.8 shows the change in AC bus voltage profile with load
changes from 2,528 to 3,792 W in five steps over a period of 6 seconds.

Table E.5   Voltage values for load change from the model for Case 2(II)

Load change	Load (W)	Actual voltage in AC bus (V)	Voltage change
Corresponding to maximum power capacity of array (3,181 W)	3,160	415	Bus voltage maintained at specified level
10% reduction	2,844	429	3.37% rise
20% reduction	2,528	443	6.74% rise
10% increase	3,476	402	3.13% drop
20% increase	3,792	390	6.02% drop

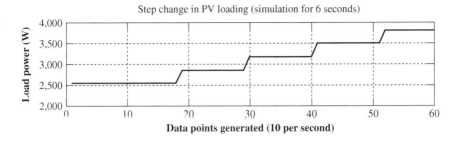

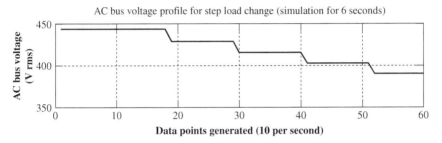

Figure E.8   AC voltage profile for step load change for E = 800 W/m² *and*
            T = 298 K

E.4.3   Case 3: Modelling of load-shedding scheme

The model has been used to develop a load-shedding scheme for the stand-alone PV system shown in Figure E.6. The load-shedding scheme is aimed at maintaining the loading level at the AC bus at a level suitable for the maximum power capacity of the PV system such that voltage at AC bus does not drop below a preset percentage. For this case, the allowable voltage drop has been chosen to be 2%. However, this limit can be changed by the user. The load-shedding module measures the voltage change in voltage profile at the 415 V AC bus and generates a trip signal (logic 0) after a time delay of 2 seconds, for opening the extra load circuit breakers for shedding these loads. The time delay of 2 seconds has been included in the programme in order to avoid spurious tripping in case of voltage transients caused by switching of other loads in the same bus. This time delay setting can also be changed by the user to suit his/her needs.

The present case shows the switching on of 20% extra load of 820 W above the maximum capacity of 4,100 W at $t = 0.2$ seconds for $T = 298$ K and $E = 1,000$ W/m² (Table E.2). The voltage sensing module endures the extra loading for 2 seconds and then generates a trip signal and sheds the extra load. The load power, voltage profile and trip signal generation are shown in Figure E.9.

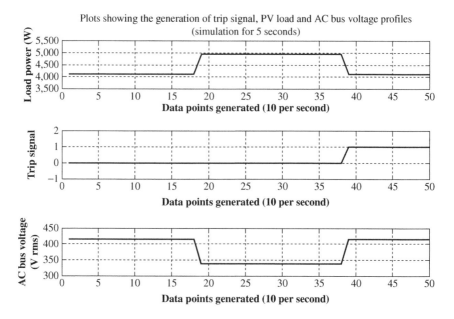

Figure E.9   Plots showing the generation of trip signal, PV loading and AC voltage profiles

Appendix F
Setting of market clearing price (MCP) in Microgrid power scenario

F.1 Proposed market structure for Microgrid participation

This section proposes and analyses a pricing mechanism for Microgrid energy in the competitive electricity market where the Microgrid CC is made to participate in the bidding process to settle the market cleaning price (MCP). Two important market settlement techniques, day-ahead and real-time, are considered with the marketing strategies for renewable distributed generator (DGs), viz. PV and wind generation. The main idea is to determine the MCP for the dispatch by an aggregate of different types of DGs to an aggregate of different types of consumers. These consumers are categorised as sheddable loads and uninterruptible loads.

An electricity market system affects the purchase and sale of electricity using supply and demand to set the price. Reducing electricity price is invariably the first reason given for introducing competitive electricity markets. Microgrid operates in a local market and usually caters to medium industrial/commercial and residential customers. These customers do not have the financial incentives and the expertise required to contribute effectively in the pricing mechanism to such a complex local market. Due to this lack of representation, most electricity markets do not treat consumers as a genuine demand-side capable of making rational decisions, but simply as a load that needs to be served under all conditions. Active participation in these markets by demand-side still remains minimal.

The main idea is to determine the MCP for the dispatch by an aggregate of different types of DGs to an aggregate of different types of consumers. These consumers are categorised as sheddable loads and uninterruptible loads. Five different generator bidders are considered, viz.:

(1) *Bidder 1* – Microturbine
(2) *Bidder 2* – Fuel cell

sales bids to satisfy all the accepted purchase bids. The sales bids are usually arranged from the lowest offer price to the highest offer price, i.e. in the bottom-up order. Purchase bids, on the contrary, are arranged from their highest offer price to the lowest offer price, i.e. in the top-down order. At the MCP, the total sales bids would be equal to the total purchase bids. In a market, both the supply and demand bids are of the same type, i.e. either block or linear bids. This section presents the detailed analysis of MCP in the competitive market for linear bid cases.

F.4.1   Single-side bid market

In this market supply companies participate in the bidding, and demand of the consumers is considered as constant irrespective of the market price. The market is considered to comprise CHP generators, renewables and diesel back-up generators. Diesel generator is generally used as back-up but for comparison purpose it has been taken as a mainstream generator.

Now, if $Q_1(p)$ kW be generated by Bidder 1 at price p \$/kWh, then supply curve can be expressed as

$$Q_1(p) = \frac{p}{m_{s1}} = Q_{1elec} + Q_{1Th} \tag{F.1}$$

where Q_{1elec} is the electrical kW generated by microturbine, Q_{1Th} the thermal energy generated by microturbine, which is converted into equivalent electrical load using Joule's constant, and m_{s1} the slope of the linear supply curve of Bidder 1.

Similarly, if $Q_2(p)$ kW be generated by Bidder 2 at price p \$/kWh, then supply curve can be expressed as

$$Q_2(p) = \frac{p}{m_{s2}} = Q_{2elec} + Q_{2Th} \tag{F.2}$$

where Q_{2elec} is the electrical kW generated by fuel cell system and Q_{2Th} the thermal energy generated by fuel cell, which is converted into equivalent electrical load using Joule's constant, and m_{s2} the slope of the linear supply curve of Bidder 2.

Likewise, combined supply curve for N bidders will be

$$\begin{aligned}
Q(p) &= Q_1(p) + Q_2(p) + \cdots + Q_N(p) \\
&= \frac{p}{m_{s1}} + \frac{p}{m_{s2}} + \cdots + \frac{p}{m_{sN}} \\
&= p \sum_{j=1}^{N} \frac{1}{m_{sj}}
\end{aligned} \tag{F.3}$$

Assuming demand is fixed at D, at MCP (p^*):

$$Q(p^*) = D$$

or

$$p^* \sum_{j=1}^{N} \frac{1}{m_{sj}} = D$$

Therefore,

$$p^* = \frac{D}{\sum_{j=1}^{N} (1/m_{sj})} \tag{F.4}$$

In (F.4), it is assumed that bidders have enough capacity of generation. If the capacity limit, minimum generation (Q_{min}) to maximum generation (Q_{max}), is specified then the combined supply curve (F.3) can be represented as

$$Q(p) = p \sum_{j=1}^{N} \frac{1}{m_{sj}} [U(Q - Q_{min}) - U(Q - Q_{max})] \tag{F.5}$$

where

$$U(Q - Q_{min}) = 1, \quad \text{when } Q \geq Q_{min}$$
$$= 0, \quad \text{when } Q < Q_{min}$$

and

$$U(Q - Q_{max}) = 1, \quad \text{when } Q \geq Q_{max}$$
$$= 0, \quad \text{when } Q < Q_{max}$$

Equating (F.5) with the demand 'D', the MCP (p^*) can be determined.

F.4.2 Double-side bid market

In this market, elasticity of demand curve has been considered. Both supply- and demand-side bidding are taken into account for determination of MCP (p^*). Both linear supply and demand variations with price have been considered for analysis.

Let $D(p)$ be the combined demand at price 'p' \$/kWh obtained from bids of N numbers of consumers participating in the market. $D(p)$ can be expressed as

$$D(p) = \sum_{j=1}^{N} \frac{p_0}{m_{dj}} - \sum_{j=1}^{N} \frac{p}{m_{dj}} \tag{F.6}$$

Figure F.2 shows the linear demand and supply bid curves. Here p_0 is the price axis intercept of demand curve that varies with type of consumers. If at a particular price (p), $D(p)$ is considered aggregated demand for all the participating consumers; therefore,

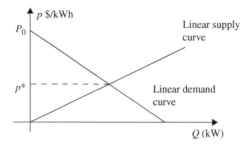

Figure F.2   Linear demand and supply bid curves

$$D(p) = \sum_{j=1}^{N} \frac{p_0}{m_{\mathrm{d}j}} - p \sum_{j=1}^{N} \frac{1}{m_{\mathrm{d}j}} \tag{F.7}$$

At the MCP (p^*),

$$p^* \sum_{j=1}^{N} \frac{1}{m_{\mathrm{s}j}} = \sum_{j=1}^{N} \frac{p_0}{m_{\mathrm{d}j}} - p^* \sum_{j=1}^{N} \frac{1}{m_{\mathrm{d}j}} \tag{F.8}$$

Therefore, MCP can be calculated from the following expression:

$$p^* = \frac{\sum_{j=1}^{N} (p_0/m_{\mathrm{d}j})}{\sum_{j=1}^{N} ((1/m_{\mathrm{s}j}) + (1/m_{\mathrm{d}j}))} \tag{F.9}$$

F.5   Case study

The Microgrid system under consideration comprises the following generators:

(1) *Bidder 1* – Microturbine
(2) *Bidder 2* – Fuel cell
(3) *Bidder 3* – Diesel generator
(4) *Bidder 4* – Wind generator
(5) *Bidder 5* – Solar PV.

These are given in Table F.1.

F.5.1   Case 1: Linear supply bid with fixed demand (i.e. single-sided bid market)

In this case, a constant demand of 80 kW is considered. Analysis is performed for the following situations:

(1) The renewable generators are considered non-available. Demand is met only by the CHP generators (Bidders 1 and 2) and diesel generator (Bidder 3). The individual and cumulative supply curves for Bidders 1, 2 and 3 are shown in Figure F.3.

Table F.1   Bidder parameters

Generators	m_s ($/kWh)	Q_{gmax} (kW)	Q_{gmin} (kW)	Heat rate (kJ/kWh)
Bidder 1 (Microturbine)	0.1056	30	Minimum power for satisfying the thermal load	12,186
Bidder 2 (Fuel cell)	0.1386	50	Do	9,480
Bidder 3 (Diesel generator)	0.063	60	0	–
Bidder 4 (Wind generator)	0.27	10	0	–
Bidder 5 (Solar PV)	0.4756	20	0	–

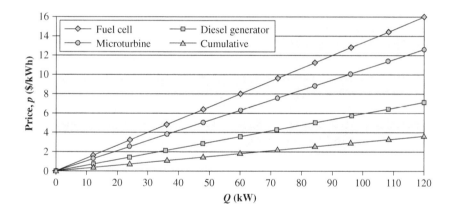

Figure F.3   Individual and combined supply curves for Bidders 1, 2 and 3

MCP is obtained from the intersection of the cumulative supply curve and the vertical line drawn from $Q = 80$ kW. For this case, MCP = $2.5/kWh.

Power supplied by each generator to meet the demand of 80 kW is shown in Table F.2.

(2) In this case, it is considered that the combined generation from the renewable generators is 30 kW and that they do not participate in the bidding process. It is also considered that both wind and PV generation are available during the daytime, which only reduces the total dispatch from Bidders 1, 2 and 3 from 80 to 50 kW. In this case, MCP is obtained from the intersection of the cumulative supply curve and the vertical line drawn from $Q = 50$ kW. For this case, MCP = $1.5/kWh. Since MCP is reduced, it may not be possible to recover the cost of renewables and also the excess cost during the

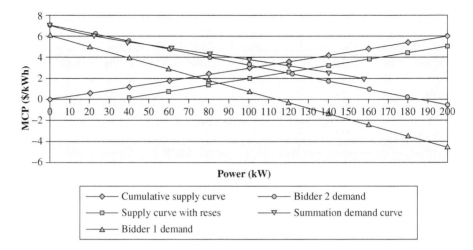

Figure F.5   Supply (cumulative) and demand (individual and cumulative) curves

Table F.4   Linear bid data

Consumer	m_{dj} ($/kWh/kW)	p_0 ($/kWh)
Bidder 1	0.041	6.0
Bidder 2	0.077	7.0

If renewables do not participate in the bidding process, then their contribution of 30 kW will reduce the MCP to \$3/kWh and consumption of the consumers will increase to

Bidder 1 = 72 kW
Bidder 2 = 52 kW
Total demand = 124 kW

From Figure F.5, when renewables supply 30 kW, then supply of other three bidders will be as follows:

Bidder 1 (Microturbine) = 23 kW
Bidder 2 (Fuel cell) = 29 kW
Bidder 3 (Diesel generator) = 49 kW
Total generation = 101 kW

From Figure F.6, when bidding rate of renewables is less than 1.0, then there is no impact on the MCP with restricted renewables of 30 kW. With the increase in bidding rate MCP increases, but power dispatched from renewables is less. Table F.5 shows the payments at various MCP and the corresponding output.

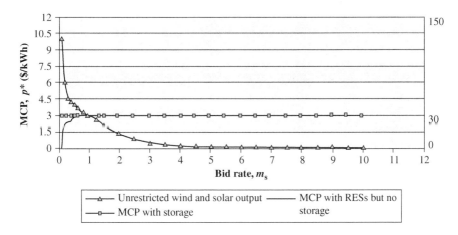

Figure F.6 MCP versus bid rate, and output versus bid rate curves when demand is elastic

Table F.5 Bidder payments and output at different MCPs

Generators	Without renewables		With renewables	
	Power (kW)	**Payment (at \$3.4/kWh)**	**Power (kW)**	**Payment (at \$3/kWh)**
Supply-side payment				
Bidder 1	33	112.2	23	69
Bidder 2	26	88.4	29	87
Bidder 3	51	173.4	49	147
Renewables	0	–	30	90
Total payment	110	374.0	131	393
Demand-side				
Bidder 1	64	217.6	76	228
Bidder 2	46	156.4	55	165
Total payment	110	374.0	131	393

F.6 Impacts on electricity market

The bidding strategies used by generating companies with the goal of max-imising their own profits show various potential possibilities to exercise market power. Market power is simply the power that market participants hold to manipulate the market in their own favour. Various reasons for the existence of market power are transmission congestion, market players and market structure.

Congestion is common in the electricity transmission system. Microgrid system itself provides respite from utility grid congestion. Scheduling of energy production in the day-ahead market will also help to mitigate transmission congestion issues. A large company (i.e. a big market player) can easily manipulate energy prices that are set far from its marginal cost. In the Microgrid market both conventional (microturbine, fuel cell) and Renewable Energy Sources (RESs) participate. Due to inconsistent behaviour of renewables, the market structure and market rules are also important causes for some kind of exercise of market power, such as what pricing mechanism is implied – uniform price or pay-as-bid. Chances of market volatility in the Microgrid market are almost absent. Though a perfect match between power production and power demand is hardly possible, still due to presence of the storage system this gap can be mitigated easily. Carbon emission alerts us every time to shift the nature of electricity generation from the fossil fuel type to non-conventional type. In this case, Microgrid system of generation has an edge over the conventional type. Carbon taxation will indirectly encourage the Microgrid system.

Appendix G
Islanding operation of distributed generators in active distribution networks – simulation studies

G.1   Background

Due to growing power demand and increasing concern about the environmental pollution caused by fossil fuelled-power plants, the distributed generation concept is gaining greater commercial and technical importance all over the world. Distributed generation encompasses the interconnection of small-scale, on-site DGs with the main utility grid at distribution voltage level. DGs constitute non-conventional and renewable energy sources like solar PV, wind turbines, fuel cells, mini/micro hydro, tidal and wave generators and microturbines. These technologies are being preferred for their high energy efficiency (microturbine or fuel cell-based CHP systems), low environmental impact (PV, wind, hydro, etc.) and their applicability as uninterruptible power supplies to PQ (power quality) sensitive loads. Electricity market reforms and advancements in electronics/communication technology are currently enabling the improved control of geographically distributed DGs through advanced SCADA (supervisory control and data acquisition). Research has been carried out on how interconnected DGs can be operated as Microgrids in both grid-connected and stand-alone modes.

Surveys indicate that in spite of increasing DG penetration, power engineers, network operators, regulators and other stakeholders are hesitant regarding incorporation of DGs into existing system and their independent operation as power islands due to various technical impacts of DG penetration on the existing utility. Power island operation is the independent operation of DGs to supply their own as well as utility loads. In this mode of operation, the DGs supply the utility loads at the utility bus and maintain full frequency and voltage controllability of that bus.

Studies show that a high degree of penetration of DG (more than 20%) as well as DG placement and sizing have considerable impact on operation, control, protection and reliability of the existing utility. These issues must be critically assessed and resolved before allowing the market participation and power island operation of DGs. This is necessary for utilising full DG potential for generation augmentation, for enhancing power quality (PQ) and reliability

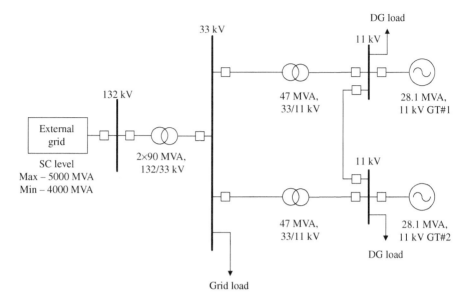

Figure G.2   System 2

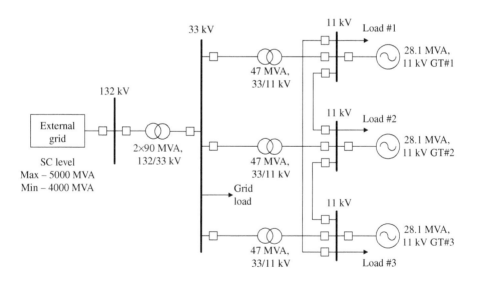

Figure G.3   System 3

the GTs are separately connected to the grid at 11 kV through 33/11 kV transformers. GT#1, GT#2 and GT#3 are connected to one another through interties as delta.

For all the systems the DGs, when grid-connected at 11 kV bus, are operated in the PQ mode and the voltage and frequency at the 11 kV bus are regulated by the grid. On islanding, the DGs are switched from PQ mode to V-f mode and the 11 kV bus voltage and frequency are then regulated by the DGs. The 132 kV grid is assumed to be of very high pool with respect to the DGs. In this study the maximum and minimum short-circuit levels of the grid are considered to be 5,000 and 4,000 MVA, respectively, and the maximum level is used for study. Grid load at 33 kV bus is taken to be 100 kW while each GT is loaded to a maximum of 23 MW on islanding. V-f controllers for DGs are modelled based on standard IEEE excitation controller and frequency controller models.

G.3 Case studies

The successful islanding operation of DGs has been showcased through several case studies for all the three system configurations described in Section G.2. For all the studies the voltages and frequencies are plotted for the 11 kV buses to which the DGs are connected.

G.3.1 Case study 1

The study is performed on System 1. Simple islanding takes place at $t = 25$ seconds. The controllers are switched on after islanding. The voltage and frequency plots in Figure G.4 show that the bus voltage takes about 7 seconds and system frequency about 6 seconds to settle after islanding. The controllers are capable to arrest the voltage and frequency excursions within permissible limits.

G.3.2 Case study 2

G.3.2.1 Case study 2(a)

The study is performed on System 2. Simple islanding takes place for both GT#1 and GT#2 at $t = 25$ seconds. The intertie between GT#1 and GT#2 is connected at the instance of islanding. In this case, both the interconnected DGs are not allowed to switch from PQ mode to V-f mode to avoid control conflict after islanding. The control strategy is to use one DG as the master controller while the other remains the slave. The master DG is switched on to V-f mode after islanding to control the overall island voltage and frequency, while the slave DG is maintained at fixed generation. On islanding any extra loading is taken up by the master DG. In this study GT#1 becomes the master DG and GT#2 the slave after islanding. The bus voltage takes about 11 seconds and system frequency takes about 10 seconds to settle after islanding. Bus voltage and frequency plots of Figure G.5 show that the bus voltage of the slave DG is slightly less than that of the master DG bus due to the drop across the interconnection tie line. The master controller is quite capable of arresting the voltage and frequency excursions of the island within permissible limits. Both the DGs are capable of becoming the master, the other being the slave. Similar responses are obtained with GT#2 as the master and GT#1 as the slave.

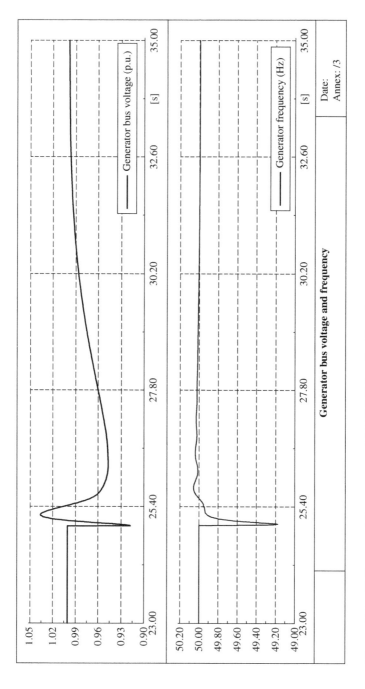

Figure G.4   Voltage and frequency at 11 kV DG bus for Case 1

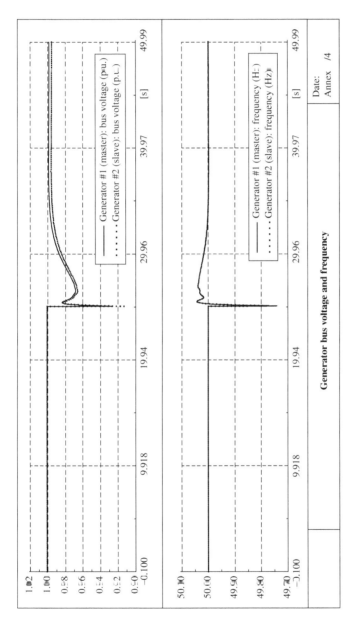

Figure G.5 Voltage and frequency at 11 kV DG bus for Case 2(a)

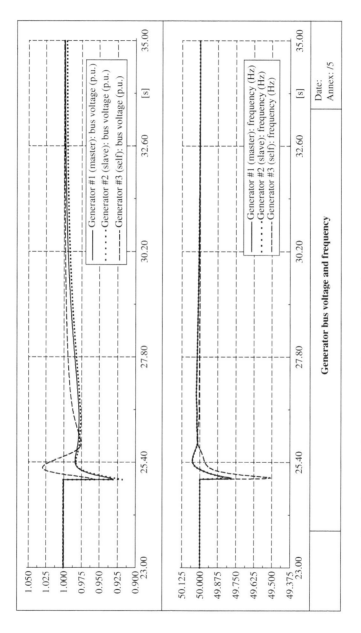

Figure G.7   Bus voltage and frequency for Case study 3(a)

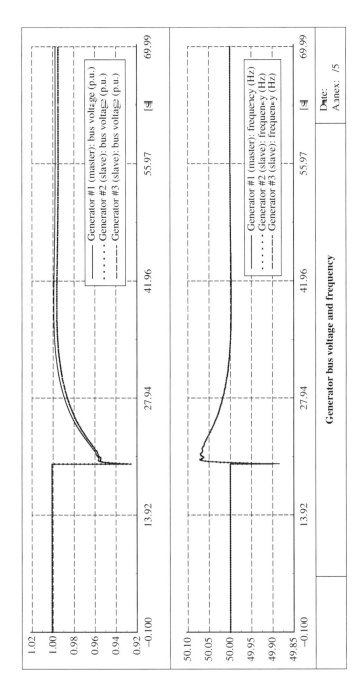

Figure G.3 Bus voltage and frequency for Case study 3(b)

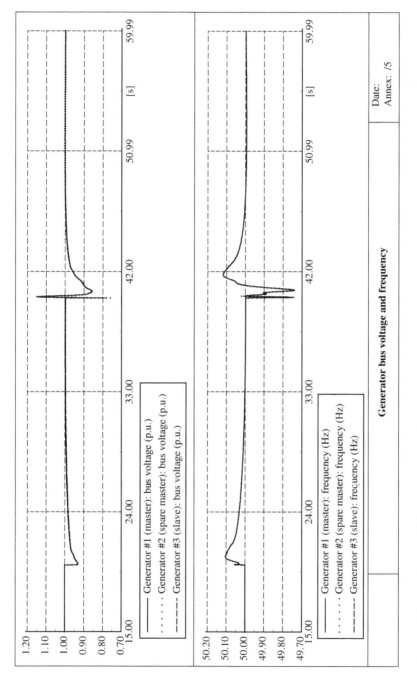

Figure G.9   Voltage and frequency at 11 kV DG bus for Case 3(c)

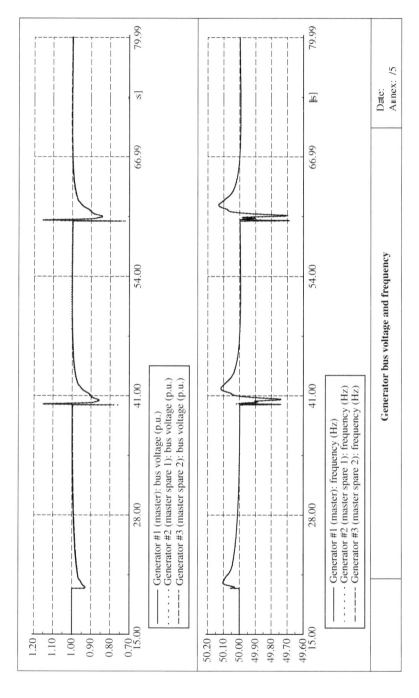

Figure G.10 Voltage and frequency at 11 kV DG bus for Case 3(d)

U.S. Department of Energy, The Electric Power Outages in The Western United States, July 23, 1996. (DOE/PO-0050), August 1996 (www.energy.gov/library/index).

Gow J.A. and Manning C.D., 'Development of a Model for Photovoltaic Arrays Suitable for Use in Simulation Studies of Solar Energy Conversion Systems', Proc. of 6th International Conference on Power Electronics and Variable Speed Drives, pp. 69–74, 1996.

King D.L., Dudley J.K. and Boyson W.E., 'PVSIM: A Simulation Program for Photovoltaic Cells, Modules, and Arrays', Proc. of 25th PVSC, Washington, D.C., May 13–17, 1996.

Beukes H., Enslin J. and Spee R., 'Busbar Design Considerations for High Power IGBT Converters', Proc. of Power Electronics Specialist Conference, Vol. 2, June 22–27, 1997.

Patterson D. and Hennessy J., *Computer Organization & Design*, Morgan Kaufmann Publishers Inc., San Francisco, CA, 1997.

Khouzam K., 'Prospect of Domestic Grid Connected PV Systems under Existing Tariff Conditions', Proc. of 26th IEEE PVSC, September 30–October 3, 1997.

National Regulatory Research Institute, NRRI Reliability Survey (internal document), 1998 (www.nrri.ohio-state.edu).

Kelley A., Harris M., Hartzell D. and Darcy D., 'Coordinated Interconnect: A Philosophical Change in the Design and Construction of Power Electronic Converters', IEEE Industry Applications Conference, Vol. 2, October 12–15, 1998.

Hopkins D., Mathuna S., Alderman A. and Flannery J., 'A Framework for Developing Power Electronics Packaging', Proc. of the Applied Power Electronics Conference and Exposition, Vol. 1, February 15–19, 1998.

CIGRE, 'Impact of Increasing Contribution of Dispersed Generation on the Power System', Final Report, Study Committee No. 37, 1998.

Moore T., 'Emerging Markets for Distributed Resources', *EPRI Journal*, Vol. 23, Issue No. 2, pp. 8–17, 1998.

Lasseter R.H., 'Control of Distributed Resources', Proc. of International Conference on Bulk Power Systems Dynamics and Control IV: Restructuring, Santorini, Greece, pp. 323–330, August 1998.

Hall D.J. and Colclaser R.G., 'Transient Modeling and Simulation of Tubular Solid Oxide Fuel Cells', *IEEE Transactions on Energy Conversion*, Vol. 14, pp. 749–753, September 1999.

Meinhardt M., Alderman A., Flannery J., Cheasty P., Eckert S. and Mathuna C., 'STATPEP—Current Status of Power Electronics Packaging for Power Supplies—Methodology', Applied Power Electronics Conference and Exposition, Vol. 1, March 14–18, 1999.

IEEE Trial Use Guide for Electric Power Distribution Reliability Indices, IEEE Std. 1366–1998, April 1999.

Neal Elliott R. and Mark Spurr M., 'Combined Heat and Power: Capturing Wasted Energy', ACEEE, May 1999 (www.aceee.org/pubs).

Van Wyk J.D. and Lee F., 'Power Electronics Technology at the Dawn of the New Millennium—Status and Future', Power Electronics Specialists Conference, Vol. 1, June 27–July 1, 1999.

Hudgins J., Mookken J., Beker B. and Dougal R., 'The New Paradigm in Power Electronics Design', IEEE Power Electronics and Drive Systems Conference, Vol. 1, July 27–29, 1999.

Hadjsaid N., Canard J.F. and Dumas F., 'Dispersed Generation Impact on Distribution Networks', *IEEE Computer Applications in Power*, Vol. 12, Issue No. 2, pp. 22–28, 1999.

Spee R. and Bhowmik S., 'Wind Turbines', *Encyclopedia of Electrical Engineering*, Wiley, New York, 1999.

2000

Jenkins N., Allan R., Crossley P.A., Kirschen D. and Strbac G., 'Embedded Generation', IEE Power and Energy Series 31, The IEE, London, UK, 2000.

Venkataramanan G., 'An Integrated Architectural Framework for Power Conversion Systems', IEEE International Workshop on Integrated Power Packaging, Waltham-Boston, MA, July 14–15, 2000.

Lee F. and Peng D., 'Power Electronics Building Block and System Integration', Proc. of Power Electronics and Motion Control Conference, Vol. 1, August 15–18, 2000.

Ericsen T., 'Power Electronic Building Blocks—A Systematic Approach to Power Electronics', Proc. of IEEE Power Engineering Society Summer Meeting, Vol. 2, July 16–20, 2000.

Kehl D. and Beihoff B., 'The Future of Electronic Packaging for Solid State Power Technology: The Transition of E-Packaging to Electromechanical Engineering', Proc. of IEEE Power Engineering Society Summer Meeting, Vol. 2, July 16–20, 2000.

Jacobsen J. and Hopkins D., 'Optimally Selecting Packaging Technologies and Circuit Partitions Based on Cost and Performance', Proc. of the Applied Power Electronics Conference and Exposition, Vol. 1, February 6–10, 2000.

Alderfer R., Brent M., Eldridge M. and Starrs T.J., 'Making Connections: Case Studies of Interconnection Barriers and their Impact on Distributed Power Projects', Golden, CO, National Renewable Energy Laboratory, May 2000 (www.eren.doe.gov/distributedpower/barriersreport).

Edison Electric Institute, 'America's Electric Utilities: Committed to Reliable Service,' May 2000 (www.eei.org).

Federal Energy Regulatory Commission, Staff Report to the Federal Energy Regulatory Commission on Western Markets and the Causes of the Summer 2000 Price Abnormalities, Part-1 of Staff Report on U.S. Bulk Power Markets, November 2000 (www.ferc.fed.us/electric/bulkpower).

U.S. Department of Energy, Final Report of the US DOE Power Outage Study Team, March 2000 (www.energy.gov/library/index).

Hurst E., 'Electric Reliability: Potential Problems and Possible Solutions', Edison Electric Institute, May 2000 (www.eei.org/issues/reliability).

National Rural Electric Cooperative Association, White Paper on Distributed Generation, 2000 (www.nreca.org).

North American Electric Reliability Council, 2000 Reliability Assessment: Reliability of the Bulk Electric Supply in North America, May 2000 (www.nerc.com/publications/annual).

U.S. Department of Energy, Interim Report of the US DOE Power Outage Study Team, January 2000 (www.energy.gov/library/index).

U.S. Department of Energy, Horizontal Market Power in Restructured Electricity Markets, March 2000 (www.energy.gov/library/index).

North American Electric Reliability Council, Reliability Assessment: 1999–2008, May 2000 (www.nerc.com/publications/annual).

Kirby B. and Kueck J., 'How Buildings Can Prosper by Interacting with Restructured Electricity Markets', ACEEE, ACEEE Summer Study on Energy Efficiency in Buildings, 2000 (www.ornl.gov/ORNL/BTC/Restructuring/pub.htm).

Ericsen T., 'Power Electronic Building Blocks—A Systematic Approach to Power Electronics', Proc. of IEEE Power Engineering Society Summer Meeting, Vol. 2, July 16–20, 2000.

Kehl D. and Beihoff B., 'The Future of Electronic Packaging for Solid State Power Technology: The Transition of E-packaging to Electromechanical Engineering', Proc. of IEEE Power Engineering Society Summer Meeting, Vol. 2, July 16–20, 2000.

Jacobsen J. and Hopkins D., 'Optimally Selecting Packaging Technologies and Circuit Partitions Based on Cost and Performance', Proc. of the Applied Power Electronics Conference and Exposition, Vol. 1, February 6–10, 2000.

Li S., Tomsovic K. and Hiyama T., 'Load Following Functions Using Distributed Energy Resources', Proc. of IEEE/PES 2000 Summer Meeting, Seattle, WA, pp. 1756–1761, July 2000.

Padulles J., Ault G.W. and McDonald J.R., 'An Approach to the Dynamic Modelling of Fuel Cell Characteristics for Distributed Generation Operation', Proc. of IEEE/PES Winter Meeting, Vol. 1, pp. 134–138, January 2000.

Padulles J., Ault G.W. and McDonald J.R., 'An Integrated SOFC Plant Dynamic Model for Power Systems Simulation', *Journal of Power Source*, Vol. 86, pp. 495–500, March 2000.

Kelber C.R. and Schumacher W., 'Adjustable Speed Constant Frequency Energy Generation with Doubly-Fed Induction Machines', Proc. of the European Conference on Variable Speed in Small Hydro, Grenoble, France, 2000.

2001

Jinghong G., Celanovic I. and Borojevic D., 'Distributed Software Architecture of PEBB-Based Plug and Play Power Electronics Systems', Proc. of IEEE Applied Power Electronics Conference and Exposition, Vol. 2, March 4–8, 2001.

Venkataramanan G., 'Integration of Distributed Technologies—Standard Power Electronic Interfaces', Report for California Energy Commission, Prepared by Power System Engineering Research Center-Wisconsin (PSERC), Wisconsin Power Electronics Research Center (WisPERC), September 2001.

Harrison J. and Redford S., 'Domestic CHP: What are the Benefits?', EA Technology Report to the Energy Saving Trust, 2001.

Cowart R., 'Efficient Reliability, The Critical Role of Demand Side Resources in Power Systems and Markets', the Regulatory Assistance Project, June 2001 (www.rapmaine.org/efficiency.html).

Eriksen P., 'Economic and Environmental Dispatch of Power/CHP Production Systems', *Electric Power Systems Research*, Vol. 57, Issue No. 1, pp. 33–39, 2001.

Lemar P.L., 'The Potential Impact of Policies to Promote Combined Heat and Power in US Industry', *Energy Policy*, Vol. 29, Issue No. 14, pp. 1243–1254, 2001.

Chapman R.E., 'How Interoperability Saves Money', *ASHRAE Journal*, Atlanta, GA, Vol. 43, February 2001 (www.ashraejournal.org).

Jinghong G., Celanovic I. and Borojevic D., 'Distributed Software Architecture of PEBB-Based Plug and Play Power Electronics Systems', Proc. of IEEE Applied Power Electronics Conference and Exposition, Vol. 2, March 4–8, 2001.

Econnect, 'Assessment of Islanded Operation of Distribution Networks and Measures for Protection', ETSU K/EL/00235/REP, 2001.

Campanari S., 'Thermodynamic Model and Parametric Analysis of a Tubular SOFC Module', *Journal of Power Sources*, Vol. 92, Issue No. 1–2, pp. 26–34, January 2001.

Thorstensen B., 'A Parametric Study of Fuel Cell System Efficiency Under Full and Part Load Operations', *Journal of Power Sources*, Vol. 92, Issue No. 1–2, pp. 9–16, January 2001.

Suter M., 'Active Filter for a Microturbine', Proc. of the IEEE Telecommunications Energy Conference, pp. 162–165, October 2001.

Nagpal M., Moshref A., Morison G.K. and Kundur P., 'Experience with Testing and Modeling of Gas Turbines', Proc. of IEEE Power Engineering Society Winter Meeting, Vol. 2, pp. 652–656, February 2001.

King W.L., Polinder H. and Slootweg J.G., 'Dynamic Modeling of a Wind Turbine with Doubly Fed Induction Generator', Proc. of IEEE Power Engineering Society Summer Meeting, Vancouver, Canada, July 2001.

Hofmann W. and Okafor F., 'Doubly-Fed Full-Controlled Induction Wind Generator for Optimal Power Utilisation', Proc. of the PEDS'01, 2001.

Iannucci J.J., Cibulka L., Eyer J.M. and Pupp R.L., 'DER Benefits Analysis Studies', Final Report, NREL/SR-620-34636, Golden, CO, National Renewable Energy Laboratory, 2003 (www.nrel.gov).

Lasseter R., Abbas A., Marnay C., Stevens J., Dagle J., Guttromson R. *et al.*, 'The CERTS Microgrid Concept', CEC Consultant Report P500-03-089F. Sacramento, CA, California Energy Commission, October 2003 (www.energy.ca.gov).

National Renewable Energy Laboratory (NREL), 'Gas-Fired Distributed Energy Resource Characterizations', NREL/TP-620-34783, Golden, CO, National Renewable Energy Laboratory, November 2003 (www.nrel.gov).

Union for the Co-ordination of Transmission of Electricity (UCTE), 'Interim Report of the Investigation Committee of the 28 September 2003 Blackout in Italy. Brussels, Belgium', UCTE, April 2003 (www.ucte.org).

U.S.–Canada Power System Outage Task Force (OTF), 'Interim Report: Causes of the August 14th Blackout in the United States and Canada', U.S. Department of Energy, November 2003 (https://reports.energy.gov).

Bailey O., Creighton C., Firestone R., Marnay C. and M. Stadler, 'Distributed Energy Resources in Practice: A Case Study Analysis and Validation of LBNL's Customer Adoption Model', Berkeley, CA, Lawrence Berkeley National Laboratory Report LBNL-52753, 2003 (www.lbl.gov).

Goldstein L., Headman B., Knowles D., Freedman S.I., Woods R. and Schweizer T., 'Gas-Fired Distributed Energy Resource Technology Characterizations', Report NREL/TP-620-34783, Golden, CO, National Renewable Energy Laboratory, 2003 (www.nrel.gov).

Wang F., Rosado S. and Boroyevich D., 'Open Modular Power Electronics Building Blocks for Utility Power System Controller Applications', Proc. of IEEE Power Electronics Specialist Conference, Vol. 4, June 15–19, 2003.

Beihoff B., 'Power Cell Concepts for the Next Generation of Appliances', Proc. of IEEE International Electric Machines and Drives Conference, Vol. 1, June 1–4, 2003.

Bertani A., Bossi C., Delfino B., Lewald N., Massucco S., Metten E. *et al.*, 'Electrical Energy Distribution Networks: Actual Situation and Perspectives for Distributed Generation', Proc. of CIRED, 17th International Conference on Electricity Distribution, Barcelona, Spain, May 2003.

Nichols D.K. and Loving K.P., 'Assessment of Microturbine Generators', Proc. of IEEE PESGM, Vol. 4, pp. 2314–2315, July 2003.

Jurado F. and Saenz J.R., 'Adaptive Control of a Fuel Cell-microturbine Hybrid Power Plant', *IEEE Transactions on Energy Conversion*, Vol. 18, Issue No. 2, pp. 342–347, June 2003.

Staunton R.H. and Ozpineci B., 'Microturbine Power Conversion Technology Review', Oak Ridge National Laboratory, Oak Ridge, TN, April 2003.

Mendez L., Narvarte L., Marsinach A.G., Izquierdo P., Carrasco L.M. and Eyras R., 'Centralized Stand Alone PV System in MicroGrid in Morocco', Proc. of 3rd World Conference on Photovoltaic Energy Conversion, Osaka, Japan, May 11–18, 2003.

Tapia A., Tapia G., Ostolaza X. and Sanez J.R., 'Modelling and Control of a Wind Turbine Driven Doubly Fed Induction Generator', *IEEE Transactions on Energy Conversion*, Vol. 18, Issue No. 2, pp. 194–204, 2003.

Song Y.H. and Wang Xi-Fan (Eds.), *Operation of Market-Oriented Power Systems*, Springer, London, 2003.

Gatta F.M., Iliceto F., Lauria S. and Masato P., 'Modelling and Computer Simulation of in Dispersed Generation in Distribution Networks. Measures to Prevent Disconnection During System Disturbances', Proc. of IEEE Power Tech Conference, Bologna, Italy, 2003.

2004

Flannery P., Venkataramanan G. and Shi B., 'Integration of Distributed Technologies—Standard Power Electronic Interfaces', Report for California Energy Commission, Prepared by Consortium for Reliability Technology Solutions (CERTS), Power System Engineering Research Center-Wisconsin (PSERC), Wisconsin Power Electronics Research Center (WisPERC), April 2004.

'The Role of Distributed Generation in Power Quality and Reliability', Final Report prepared for New York State Energy Research and Development Authority, Albany, NY, Prepared by Energy and Environmental Analysis, Inc., Arlington, VA and Pace Energy Project White Plains, NY, June 2004.

Second Annual Report of the Distributed Generation Co-ordinating Group, DTI, March 2004 (www.distributed-generation.gov.uk).

Firestone R. and Marnay C., 'Energy Manager Design for Microgrids', Berkeley Laboratory Report in Draft, Berkeley, CA, Ernest Orlando Lawrence Berkeley National Laboratory, 2004 (www.lbl.gov).

Illindala M.S., Piagi P., Zhang H., Venkataramanan G. and Lasseter R.H., 'Hardware Development of a Laboratory-Scale Microgrid Phase 2: Operation and Control Two-Inverter Microgrid', NREL Report No. SR-560-35059, Golden, CO, National Renewable Energy Laboratory, 2004 (www.nrel.gov).

Lasseter R.H. and Paigi P., 'Microgrid: A Conceptual Solution', Proc. of Power Electronics Specialists Conference (PESC), Aachen, Germany, Vol. 6, pp. 4285–4290, June 2004.

Sedghisigarchi K. and Feliachi A., 'Dynamic and Transient Analysis of Power Distribution Systems with Fuel Cells—Part I: Fuel-Cell Dynamic Model', *IEEE Transactions on Energy Conversion*, Vol. 19, Issue No. 2, pp. 423–428, June 2004.

Sedghisigarchi K. and Feliachi A., 'Dynamic and Transient Analysis of Power Distribution Systems with Fuel Cells—Part II: Control and Stability Enhancement', *IEEE Transactions on Energy Conversion*, Vol. 19, Issue No. 2, pp. 429–434, June 2004.

Sakhare A.R., Davari A. and Feliachi A., 'Control of Solid Oxide Fuel Cell for Stand-Alone and Grid Connection Using Fuzzy Logic', Proc. of IEEE Thirty Sixth Southeastern Symposium, System Theory, pp. 551–555, 2004.

Jurado F., Jose R. and Fernandez S.L., 'Modeling Fuel Cell Plants on the Distribution System Using Identification Algorithms', Proc. of IEEE Electrotechnical Conference, Melecon, Vol. 3, pp. 1003–1006, May 2004.

Jurado F., Jose R. and Fernandez S.L., 'Development of the Solid Oxide Fuel Cell', *Energy Sources*, Vol. 26, Issue No. 2, pp. 177–188, February 2004.

Miao Z., Choudhry M.A., Klein R.L. and Fan L., 'Study of a Fuel Cell Power Plant in Power Distribution System—Part I: Dynamic Model', Proc. of IEEE/PES General Meeting, Vol. 2, pp. 2220–2225, June 2004.

Miao Z., Choudhry M.A., Klein R.L. and Fan L., 'Study of a Fuel Cell Power Plant in Power Distribution System—Part II: Stability Control', Proc. of IEEE/PES General Meeting, Vol. 2, pp. 1–6, June 2004.

Mazumder S.K., Acharya K., Haynes C.L., Williams R., Jr., Spakovsky M.R., Nelson D.J. *et al.*, 'Solid-Oxide-Fuel-Cell Performance and Durability: Resolution of the Effects of Power-Conditioning Systems and Application Loads', *IEEE Transactions on Power Electronics*, Vol. 19, Issue No. 5, pp. 1263–1278, September 2004.

Hatziargyriou N., Kariniotakis G., Jenkins N., Pecas Lopes J., Oyarzabal J., Kanellos F. *et al.*, 'Modelling of Microsources for Security Studies', Proc. of CIGRE Session, Paris, France, August–September 2004.

Brenna M., Tironi E. and Ubezio G., 'Proposal of a Local dc Distribution Network with Distributed Energy Resources', Proc. of International Conference on Harmonics and Quality of Power, pp. 397–402, September 2004.

Peirs J., Reynaerts D. and Verplaetsen F., 'A Microturbine for Electric Power Generation', *Sensors and Actuators, A*, Vol. 113, Issue No. 1, pp. 86–93, 2004.

Ho J.C., Chua K.J. and Chou S.K., 'Performance Study of a Microturbine System for Cogeneration Application', *Renewable Energy*, Vol. 29, Issue No. 7, pp. 1121–1133, 2004.

Bertani A., Bossi C., Fornari F., Massucco S., Spelta S. and Tivegna F., 'A Microturbine Generation System for Grid Connected and Islanding Operation', Proc. of IEEE PES Power Systems Conference and Exposition, Vol. 1, pp. 360–365, October 2004.

Mihet-Popa L. and Blaabrierg F., 'Wind Turbine Generator Modeling and Simulation Where Rotational Speed is the Controlled Variable', *IEEE Transactions on Industry Applications*, Vol. 40, Issue No. 1, January/February 2004.

Kumpulainen L. and Kauhaniemi K., 'Distributed Generation and Reclosing Coordination,' Proc. of Nordic Distribution and Asset Management Conference, NORDAC, Espoo, 2004.

2005

Li Y.H., Choi S.S. and Rajakaruna, S., 'An Analysis of the Control and Operation of a Solid Oxide Fuel-Cell Power Plant in an Isolated System', *IEEE Transactions on Energy Conversion*, Vol. 20, Issue No. 2, pp. 381–387, June 2005.

Econnect, 'Islanded Operation of Distribution Networks', DG/CG/00026/00/00, 2005.

Guda S.R., Wang C. and Nehrir M.H., 'A Simulink-Based Microturbine Model for Distributed Generation Studies', Proc. of IEEE Power Symposium, pp. 269–274, October 2005.

Wies R.W., Johnson R.A., Agrawal A.N. and Chubb T.J., 'Simulink Model for Economic Analysis and Environmental Impacts of a PV with Diesel-Battery System for Remote Villages', *IEEE Transactions on Power Systems*, Vol. 20, Issue No. 2, pp. 692–700, May 2005.

Robitaille M., Agbossou K. and Doumbia M.L., 'Modeling of an Islanding Protection Method for a Hybrid Renewable Distributed Generator', Proc. of Electrical and Computer Engineering, Canada, pp. 1477–1481, 2005.

2006

Yixiang S. and Ningsheng C., 'A General Mechanistic Model of Solid Oxide Fuel Cells', *Tsinghua Science and Technology*, Vol. 11, Issue No. 6, pp. 701–711, December 2006.

Zhang X., Li J. and Feng Z., 'Development of Control Oriented Model for the Solid Oxide Fuel Cell', *Journal of Power Source*, Vol. 160, Issue No. 1, pp. 259–267, September 2006.

Huo H.B., Zhu X.J. and Cao G.Y., 'Nonlinear Modeling of a SOFC Stack Based on a Least Squares Support Vector Machine', *Journal of Power Sources*, Vol. 162, Issue No. 2, pp. 1220–1225, November 2006.

Goel A., Mishra S. and Jha A.N., 'Power Flow Control of a Solid Oxide Fuel-Cell for Grid Connected Operation', Proc. of International Conference on Power Electronics, Drives and Energy Systems, PEDES, pp. 1–5, 2006.

Boccaletti C., Duni G., Fabbri G. and Santini E., 'Simulations Models of Fuel Cell Systems', Proc. of ICEM, Electrical Machines, Chania, Greece, September 2006.

Esmaili R., Das D., Klapp D.A., Dernici O. and Nichols D.K., 'A Novel Power Conversion System for Distributed Energy Resources', Proc. of IEEE PES General Meeting, pp. 1–6, June 2006.

Gaonkar D.N. and Patel R.N., 'Modeling and Simulation of Microturbine Based Distributed Generation System', Proc. of IEEE Power India Conference, April 2006.

Gaonkar D.N., Patel R.N. and Pillai G.N., 'Dynamic Model of Microturbine Generation System for Grid Connected/Islanding Operation', Proc. of IEEE

CIRED Seminar 2008, Smartgrids for Distribution, Frankfurt, Germany, June 23–24, 2008.

Saha A.K., Chowdhury S.P., Chowdhury S. and Crossley P.A., 'Microturbine Based Distributed Generator in Smart Grid Application', Proc. of Power Engineering Society General Meeting, Pittsburgh, PA, July 20–24, 2008.

Chowdhury S.P., Chowdhury S., Ten C.F. and Crossley P.A., 'Islanding Operation of Distributed Generators in Active Distribution Networks', Proc. of the 43rd International Universities Power Engineering Conference, Padova, Italy, September 1–4, 2008.

Saha A.K., Chowdhury S.P., Chowdhury S. and Crossley P.A., 'Study of Microturbine Models in Islanded and Grid-Connected Mode', Proc. of the 43rd International Universities Power Engineering Conference, Padova, Italy, September 1–4, 2008.

Websites

Panint Electronic Limited, Website: www.panint.com

Semikron, Website: www.semikron.com

Infineon Technologies, Website: www.infineon.com

International Recifier, Website: www.irf.com

Office of Naval Research Advance Electrical Power Systems PEBB, Website: https://aeps.onr.navy.mil

Planar Magnetics Ltd., Website: www.planarmagnetics.com

Payton Group International, Website: www.paytongroup.com

Bussco, A division of Circuit Components LLC, Website: www.bussco.com

Eldre Corporation, Website: www.busbar.com

MiniCircuits, Website: www.minicircuits.com

Assessing Product Reliability, handbook by NIST, Website: www.itl.nist.gov/div898/handbook

Switch reliability report, International Rectifier, Website: www.irf.com/productinfo/reliability

William J. Vigrass, 'Calculation of semiconductor failure rates', Website: http://rel.intersil.com/docs/rel/calculation_of_semiconductor_failure_rates.pdf

www.weibull.com/LifeDataWeb/a_brief_introduction_to_reliability.htm

Aluminum-Electrolytic Capacitor Application Guide, Website: www.cornell-dubilier.com/appguide.pdf

www.r-theta.com/products/aquasink/aquasink.pdf

Distributed Generation Coordination Group, Website: www.distributed-generation.org.uk

Charging Principles DTI/OFGEM Embedded Generation Working Group WP4, Website: www.distributed-generation.org.uk

Options for Domestic and Other Micro-Scale Generation DTI/OFGEM Embedded Generation Working Group WP5, Website: www.distributed-generation.org.uk

Industry data, Website: www.ibiblio.org/pub/academic/environment/alternative-energy/energyresources

Future Network Design, Management and Business Environment Generation DTI/OFGEM Embedded Generation Working Group WP7, Website: www.distributed-generation.org.uk

Tyndall Centre Technical Reports, Website: www.tyndall.ac.uk/publications/tech_reports

Douglas Decker, 'Federal Government: Building Management Systems/Utility Monitoring and Control Systems' Johnson Controls, Website: www.johnsoncontrols.com/cg-gov/art-utilitydereg.html

Heller Jim, 'Load Shifting with Thermal Energy Storage', Navy ENews 95b, Naval Facilities Engineering Service Center, Website: www.nfesc.navy.mil

Douglas Hinrichs D., Ray McGowan R. and Susan Conbere S., 'Integrated CHP Offers Efficiency Gains to Buildings Market', D&R International Distributed Energy Resources Team, September 2001, Energy User News, 9/27/01, Website: www.energyusernews.com

Index

Printed in the USA
CPSIA information can be obtained
at www.ICGtesting.com
JSHW011518221024
72172JS00008B/62